Dynamic Laser Speckle and Applications

OPTICAL SCIENCE AND ENGINEERING

Founding Editor
Brian J. Thompson
University of Rochester
Rochester, New York

Dynamic Laser Speckle and Applications

Edited by
Hector J. Rabal
Roberto A. Braga Jr.

CRC Press
Taylor & Francis Group
Boca Raton London New York

CRC Press is an imprint of the
Taylor & Francis Group, an **informa** business

CRC Press
Taylor & Francis Group
6000 Broken Sound Parkway NW, Suite 300
Boca Raton, FL 33487-2742

First issued in paperback 2019

ISBN-13: 978-1-4200-6015-7 (hbk)
ISBN-13: 978-0-367-38638-2 (pbk)

Library of Congress Cataloging-in-Publication Data

Dynamic laser speckle and applications / editors, Hector J. Rabal and Roberto A.
    Braga, Jr.
        p. cm. -- (Optical science and engineering ; 139)
    Includes bibliographical references and index.
    ISBN 978-1-4200-6015-7 (alk. paper)
    1. Laser speckle. I. Rabal, Hector J. II. Braga, Roberto A., Jr. III. Title. IV.
Series.

TA1677.D88 2009
621.36'6--dc22   2008040947

Visit the Taylor & Francis Web site at
http://www.taylorandfrancis.com

and the CRC Press Web site at
http://www.crcpress.com

To Our Families

Contents

Preface

Laser history may be written featuring some landmarks that represent our efforts to understand and develop this widely applicable tool. The first one can be found in the work of Albert Einstein, when in the beginning of the twentieth century he presented the theoretical basis required to obtain amplification of light in a desired frequency. The transition from theory to practice required much energy, and 40 years later materialized in the second major advance with Maimann's work in the 1950s, which triggered a new stage of research. The developments in laser sets and their applications started an avalanche similar to that occurring with electrons transitions in lasers, promoting further cooperation among researchers in physics and the applied sciences. A new laser field allowed new applications, and other uses were found, prompting more laser sets research.

Further landmarks have appeared since then, and now, in less time, new applications have been developed, along with advanced laser technologies. The research accelerated with the improvement of computers, frame grabbers, and, of course, with their price reduction.

Some research branches that have gained prominence are holography and speckle interferometry, both analogical and digital. Holography itself has provided many metrological tools for the nondestructive testing of actual applicative utility. It also called the attention of experimenters to the granular appearance of the objects that were illuminated by the coherent source.

Initially a negative in the field of holographic imagery, the speckle phenomenon soon proved itself as an aid in the development of metrological tools that peaked with the digital speckle pattern interferometry, digital holography, and optical coherence tomography.

Forrester, Cummins, Bergé, Abramson, Briers, and many others showed the way for the applications discussed in this book, demonstrating interesting uses of laser light images in biological tissue. In this direction, other researchers pursued applications in the biological material field, using the dynamic speckle laser phenomenon as an instrument for biological activity characterization. This activity was then observed as a sample under laser light, producing the dynamic speckle phenomenon, known also as biospeckle. We found dynamic speckle phenomenon to be a special way to see objects in another dimension, providing more information.

Biospeckle study constitutes a multidisciplinary area that challenges scientists to connect with one another's specialties. There is a naturally high complexity in the interaction of the laser beam with distinct biological samples, all containing a large amount of variables interacting with light. Biological tissues, in the variability of their samples, offer an even wider range for study.

This book intends to contribute by presenting some bridges created to help the work in the biological and nonbiological metrology using dynamic speckle laser. It is divided into three parts. The first is related to the presentation of the theoretical basis of the phenomenon—speckle in Chapter 1 and biospeckle in Chapter 2. The first part is concluded with the discussion in Chapter 3 on models, theoretical and numerical, to simulate the laser light interaction with the biological samples discussed. The second part of the book uses four chapters to present the main methodologies for analyzing biospeckle phenomena. Chapter 4 presents the statistical analysis, Chapter 5 the image analysis, and Chapter 6 deals with the concepts of frequency decomposition. The second part concludes in Chapter 7 with the discussion of fuzzy methods to treat dynamic speckle data. The third part of the book is dedicated to application approaches, presenting some interesting cases, especially related to agricultural areas. The examples of biospeckle metrology were divided into biological (in Chapter 8) and nonbiological materials (Chapter 9). Finishing the third part we decided to share with students and teachers, in particular, some software to facilitate easy start-up in dynamic speckle usage, which has been presented in Chapter 10.

New solutions and prospects are expected to come from interdisciplinary cross talk, and this book represents an effort in that direction with its groups of scientists communicating here their experiences and evaluating the state of the art. It is their hope to offer to the worldwide community of scientists, teachers, students, and professionals information that could be useful to further their studies and applications.

Finally, we discuss in Chapter 11 the future of the technique, presenting some challenges, introducing new research areas, and evaluating the reliability of the technique to improve biological metrology.

Considerable effort went into this book's gathering together of assessments in research so that a clear presentation could be made of the advances in the field. Generous help and excellent contributions were made by our colleagues in this regard. We especially want to offer this final version first to our colleague Nestor Gaggioli, who had a stroke at the end of the writing process. We hope to share with him, in good health, the satisfaction of finishing this challenge.

We also would like to share our gratitude with these other colleagues: Claudia Gonzalez, Nelly Cap, Christian Weber, Prof. Guillermo Kaufmann, Prof. Giovanni Rabelo, and, in particular, the people of the Taylor & Francis Group, represented by Jessica Vakili.

In Argentina, we thank the support of Consejo Nacional de Investigaciones Científicas y Técnicas, Comisión de Investigaciones Científicas de la Provincia de Buenos Aires, Universidad Nacional de La Plata, Argentina, and, in Brazil, the Universidade Federal de Lavras, CAPES, CNPq, FAPEMIG, and FINEP.

Finally, we extend our appreciation to the Third World Academy of Sciences (TWAS), International Centre of Theoretical Physics, and International Commission for Optics (ICO).

Hector J. Rabal

Roberto A. Braga Jr.

Contributors List

Ana Lucía Dai Pra
Universidad Nacional de Mar del Plata
Laboratorio de Inteligencia Artificial–
 Facultad de Ingeniería
Mar del Plata, Argentina

Antônio Tavares da Costa Jr.
Instituto de Física
Universidade Federal Fluminense
Campus da Praia Vermelha
Niterói, Brazil

Constancio Miguel Arizmendi
Departamento de Física–Facultad de
 Ingeniería
Universidad Nacional de Mar del Plata
Mar del Plata, Argentina

Elsa Noemí Hogert
Comisión Nacional de Energía
 Atómica
San Martín, Pcia. de Bs. As.
Argentina

Flávio M. Borém
Campus UFLA Dep. Engenharia
Universidade Federal de Lavras
Lavras, Brazil

Giovanni F. Rabelo
Campus UFLA Dep. Engenharia
Universidade Federal de Lavras
Lavras, Brazil

Gonzalo Hernán Sendra
Consejo Nacional de Investigaciones
 Científicas y Técnicas
Centro de Investigaciones Ópticas.
 Ciop.
Buenos Aires, Argentina

Hector Jorge Rabal
Centro de Investigaciones Ópticas
CONICET y Facultad de Ingenieria,
 UID OPTIMO
Universidad Nacional de La Plata
La Plata, Argentina

Inácio Maria Dal Fabbro
Universidade Estadual de Campinas
FEAGRI–UNICAMP–Cidade
 Universitária Zeferino Vaz
Campinas, Brazil

Joao Bosco Barreto Filho
Campus UFLA Dep. Medicina Veterinária
Universidade Federal de Lavras
Lavras

Joelma Pereira
Campus UFLA Dep. Engenharia
Universidade Federal de Lavras
Lavras, Brazil

Juan Antonio Pomarico
Universidad Nacional del Centro de la
 Província de Buenos Aires UNCPBA
Buenos Aires, Argentina

Lucía Isabel Passoni
Laboratorio de Bioingeniería–Facultad
 de Ingeniería
Universidad Nacional de Mar del Plata
Mar del Plata, Argentina

Marcelo Trivi
Centro de Investigaciones Opticas
 (CONICET–CIC)
UID Optimo, Dpto. Ciencias Básicas,
 Facultad de Ingeniería
Universidad Nacional de La Plata
La Plata, Argentina, Brazil

María Fernanda Ruiz Gale
Comisión Nacional de Energía Atómica
San Martín, Pcia. de Bs. As.
Argentina

Marlon Marcon
Campus UFLA Dep. Engenharia
Universidade Federal de Lavras
Lavras, Brazil

Mikiya Muramatsu
Instituto de Física–Universidade de São
  Paulo
São Paulo, Brazil

Néstor Gustavo Gaggioli
Optics and Laser Lab
Department ENDE–CNEA
Consejo Nacional de Investigación
  Científica y Técnica
San Martín, Pcia. de Bs. As.
Argentina

Ricardo Arizaga
Centro de Investigaciones Opticas
CONICET y Faculdad de Ingenieria,
  UID OPTIMO
Universidad Nacional de La Plata, UNLP
Argentina

Roberto A. Braga Jr.
Campus UFLA Dep. Engenharia
Universidade Federal de Lavras
Lavras, Brazil

1 The Speckle Phenomenon

Néstor Gustavo Gaggioli and
Juan Antonio Pomarico

CONTENTS

1.1   INTRODUCTION

When an optically rough surface is illuminated with light having a high degree of coherence, such as one coming from a laser, the scattered light presents a particular intensity distribution, making the surface appear to be covered with a fine granular structure. This structure, which consists of alternately dark and bright spots of variable shapes, and distributed in a random way, as shown in Figure 1.1, has no obvious relation to the macroscopic properties of the surface. Such intensity distribution, which is also observable when coherent light propagates through a medium presenting random variations in its refractive index, is known as a *speckle pattern*.

It is not mandatory to use an image forming system, such as the eye or a photographic camera, to observe a speckle pattern. Speckle patterns fill the space surrounding the diffusive medium, and so they can be registered by properly placing a photographic plate or a CCD detector at a certain distance from the object. It is usual to call speckle patterns obtained from free space propagation of light *objective speckles*, whereas those patterns obtained under image formation conditions are referred to as *subjective speckles*.

The speckle results from the temporally stationary interference of light scattered by the assembly of many optical inhomogeneities distributed randomly in space, such as a diffuse surface. In this way, multiple scattered coherent wavelets having

1

FIGURE 1.1 Picture of an actual speckle diagram.

randomly distributed amplitudes and phases come together at a given observation point in space and produce the intensity distribution as shown in Figure 1.1. The shape of the envelope of the illuminated area and the typical size of the individual speckle grains at the observation plane are determined by the detailed structure of the surface and by the aperture of the optical system, respectively. This is a point to be addressed later on in this chapter.

Since the invention of the laser in the early years of the 1960s, there has been renewed interest in the speckle, and many applications have been proposed. In this modern era of interest in the phenomenon, an unpublished report by Goodman[1] is probably still the clearest description of the basic statistics of (Gaussian) speckle. Other related works have made more contributions to this subject.[2-6]

The optical properties of laser light have brought about a renaissance in optics. In that process, considerable progress has been made in the diffraction theory of image formation, and this field has received the name of *coherent optics*. Researchers have used these results to give an interpretation to the speckle phenomenon.

In the case of image formation of a rough object using coherent light, the entrance pupil is illuminated by a speckled diffraction pattern. Such random illumination then appears at the exit pupil and so the speckling in the image has also been studied considering the exit pupil to act as a rough surface.[7-10] Some other researchers have observed and studied the effect of the degree of coherence upon the particular characteristics of speckle patterns.[11-16]

As mentioned, the random phase differences between interfering waves from the scattering centers must remain stationary for speckles to be observed. If the diffuse object is set into slow motion, such phases as well as the object region contributing to an observed disturbance of light change with time. Therefore, the speckles are observed to vary gradually. The change in phase differences causes the speckles to translate, whereas the one in the contributing region modifies the detailed structure of the pattern. This suggests that speckle fields can be used as random carriers of information and thus be employed to quantify translations, rotations, phase changes, etc., of a given rough or translucent object. Sporton[17] was probably the first to quantitatively study the translation of speckles in the diffraction field for tilt, in-plane translation, and in-plane rotation of the object, thus starting a new research field in optical metrology.

Before going into further details concerning the physical properties of speckle patterns, we will devote some paragraphs to a brief historical review of such structures.

1.2   SOME HISTORICAL CONSIDERATIONS

There is no doubt about the fact that speckle patterns attracted major interest after the first gas laser appeared on the scene. Since then, several studies have been done to explain the effects that occur when light from a laser is reflected or transmitted from a motionless rough surface. Rigden and Gordon,[18] and Oliver[19] have photographed and discussed the causes of a phenomenon in which the area illuminated by a laser beam appears to possess a remarkable granular nature not visible in ordinary light. A similar phenomenon has been noted by Langmuir.[20]

In the decade of 1960s, *speckle pattern* was the name given to the intensity pattern produced by any diffusing object illuminated by a highly coherent light beam. However, such random intensity patterns, which nowadays are invariably observed when lasers are used in the laboratory, are hardly new phenomenon. In the middle of the nineteenth century, considerable interest was shown in interference phenomena in scattered light. Good examples are Newton's diffusion rings or Quételet's[21] fringes, and also Fraunhofer's diffraction rings, which are produced when a glass plate covered with small particles diffracts fairly coherent light.

Sir Isaac Newton clearly described such random effect as it occurs at the focus of (Earth-based) telescopes;[22] he explained the trembling points by the action of atmospheric movement. At that time, many scientists supposed that, when there was a large number, n, of similar particles randomly distributed, the resulting diffraction intensity was n times that of a single particle. Lord Rayleigh,[23] however, showed that this was far from being the case, and that the resultant in any particular trial may be anywhere between 0 and n^2. After Newton and Lord Rayleigh, the earliest observation of the phenomenon of speckling appears to have been made by Exner[24]

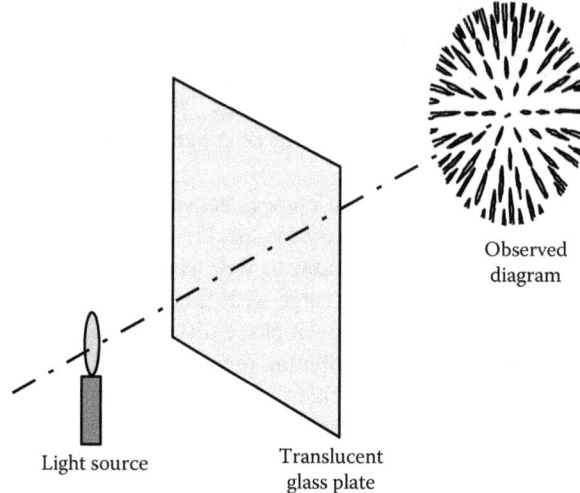

Observed
diagram

Light source   Translucent
glass plate

FIGURE 1.2 Schematic representation of the experimental setup used by Exner showing the pattern with the radial structure he could observe.

nearly 130 years ago in connection with a study of the Fraunhofer rings, formed when a beam of coherent light is diffracted by a number of particles of the same size randomly distributed. Figure 1.2 is a sketch of Exner's drawing of the radial granular structure he observed in the diffraction pattern of a glass plate on which he had breathed, under illumination by a candle.

Other significant contributions to the theory of speckle formation were also made by von Laue[25] and de Haas,[26] all prior to 1920. Von Laue obtained the diffraction pattern produced by a glass plate covered with lycopodium powder. He used light from a carbon lamp passed through a prism, selecting the region between 420 nm and 430 nm to illuminate the sample, and he also observed a pattern similar to that of Figure 1.2. However, he could not correctly explain this radial structure of the speckle. The point was correctly addressed by Schiffner[27,28] some decades later. He showed that the sizes of one irregularity in the granularity of the diffraction pattern in the radial and tangential direction are different because they depend on different ways on the diffraction angle. This is clearly observed in Figure 1.2; as the diffraction pattern extends from the center to the border, the spots in the granularity become more longish. In the 1950s, in a very interesting work, Ratcliffe[29] adapted some aspects of many-particles diffraction theory to radio waves. Subsequently, Allen and Jones[30] explained some characteristics of speckle phenomenon using the Ratcliffe results.

Some of these works have been reviewed by Schiffner[27] and Hariharan.[31] The latter made an interesting contribution, mentioning some works barely known before, such as the one by Raman using many particles with strictly monochromatic light,[32] a carefully controlled von Laue experiment made by Buchwald,[33] and the work of Ramachandran[34] in connection with the study of coronas.

1.3 PHYSICAL PROPERTIES OF SPECKLE PATTERNS

1.3.1 STATISTICS OF THE INTENSITY

Simple observation of a speckle diagram reveals that dark and bright points occur with different frequency. In fact, it can be readily seen that dark spots are much more common than bright ones. The explicit calculations leading to this conclusion can be found with a great level of details in some references.[4,5] Instead of going too deep into mathematics, we outline here the main assumptions and steps to arrive at the description of the intensity statistics of a speckle diagram. We will assume that a collimated coherent field of wavelength λ illuminates an optically rough surface; that means the chosen wavelength is much smaller than the surface height variations with respect to its mean. In practice, this is the case for most materials having optical wavelengths about 0.5 μm. Because the height of the surface varies in a random way, the spherical wavelets reflected by the surface also present a random distribution of phases (Figure 1.3a). All these wavelets interfere at a given point in space, $P(x,y,z)$. A completely analogous situation results from considering a ground glass that is being illuminated from its back by a coherent collimated beam, as shown in Figure 1.3b. The surface at which light enters the diffuser can be considered flat, and the small local curvatures at the exit surface spread the field into spherical wavelets. Additionally, the random thickness variations of the glass produce the random phases of the secondary waves.

Moreover, if an optical system is used to image the surface, diffraction considerations cannot be avoided. In fact, if speckle is to be observed with such configuration, the optical system must be properly chosen to have a point spread function broad enough to guarantee that many individual regions of the object overlap at the image plane, as sketched in Figure 1.3c.

No matter which picture is assumed, it is easily realized that the light field at a specific point $P(x,y,z)$ in a speckle pattern must be the sum of a large number N of components representing the contribution from all points on the scattering surface. Under illumination by monochromatic and fully polarized light, the contribution to the field at P produced by any surface element, j, is given by

$$u_j(P) = |u_j| e^{i\phi_j} = |u_j| e^{ikr_j} \tag{1.1}$$

r_j being the (random varying) distance from the j-th scattering surface element to the point P. The complex amplitude of the scattered field at point P can therefore be written as

$$U(P) = \frac{1}{\sqrt{N}} \sum_{j=1}^{N} u_j(P) = \frac{1}{\sqrt{N}} \sum_{j=1}^{N} |u_j| e^{i\phi_j} = \frac{1}{\sqrt{N}} \sum_{j=1}^{N} |u_j| e^{ikr_j} \tag{1.2}$$

The summation in Equation 1.2 can be considered as a random walk in the complex plane due to the random phases $\phi_j = kr_j$. This erratic motion of the field components is represented in Figure 1.4 for a few of them. Assuming that (1) the amplitude u_j and

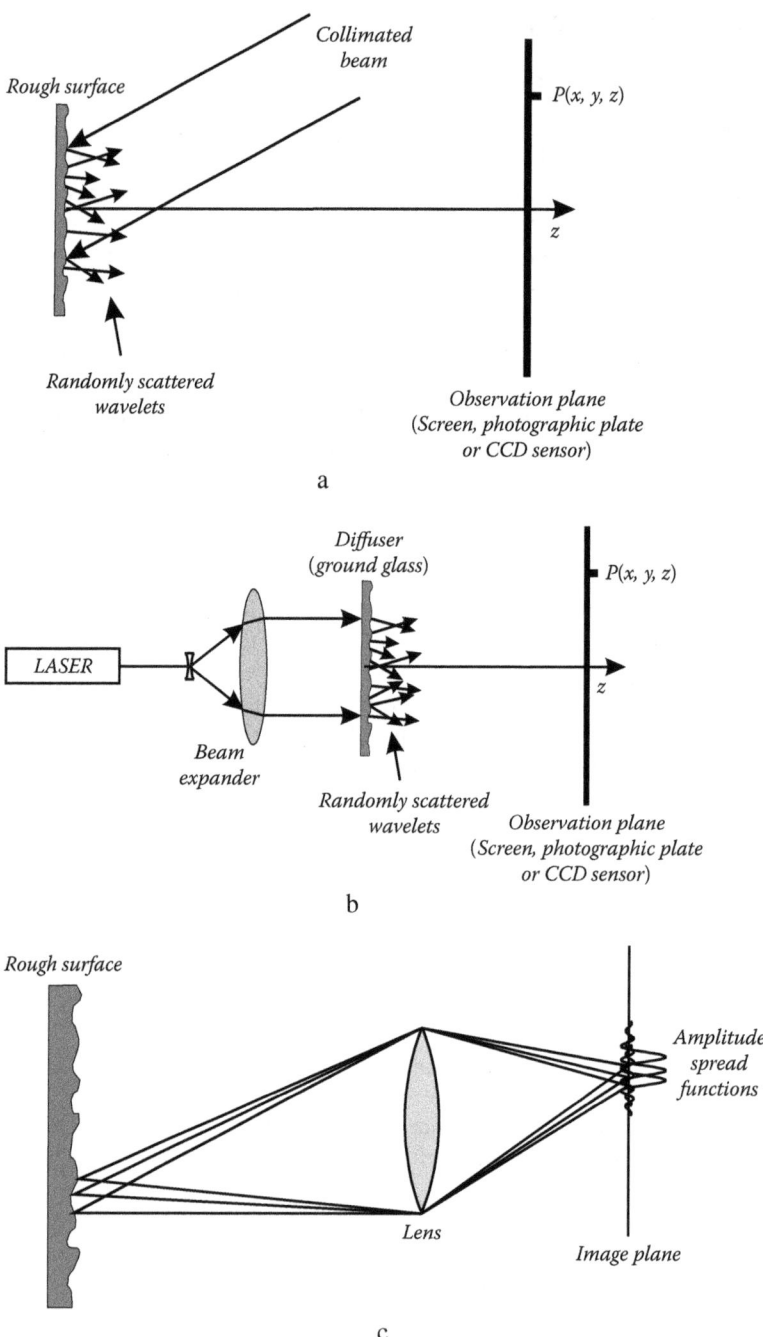

FIGURE 1.3 The physical origin of speckle patterns. (a) Diffuse reflection of coherent light from a rough surface, (b) Transmission of coherent light through a translucent object, and (c) image formation of a rough surface.

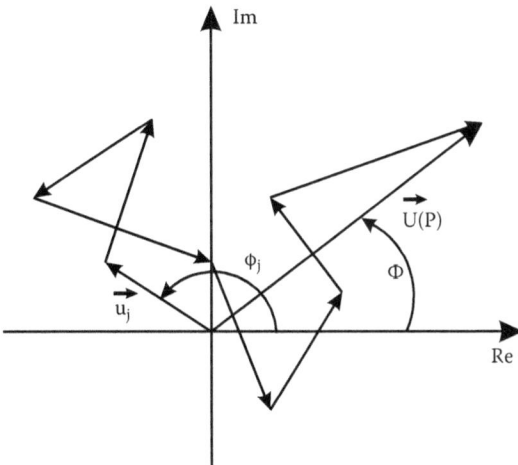

FIGURE 1.4 Several scattered fields $u_j(P)$, plotted in the complex plane with their respective random phases (ϕ_j), contributing to the total field at point P, $U(P)$.

phase ϕ_j of each field component are statistically independent of each other and are also independent of the amplitude and phase of all other field components and (2) the phases ϕ_j are uniformly distributed on the interval $(-\pi, \pi)$, which means that the surface is rough in comparison to the wavelength, and with the additional hypothesis that the number of total scattering centers N is very large, thus ensuring validity of the central limit theorem, Goodman has demonstrated that the real and imaginary parts of the resultant field are asymptotically Gaussian.[2-3] The joint probability density function for them is thus given by

$$p_{r,i}(U^{(r)}, U^{(i)}) = \frac{1}{2\pi\sigma^2} \exp\left[\frac{(U^{(r)})^2 + (U^{(i)})^2}{2\sigma^2}\right] \tag{1.3}$$

known as a *circular Gaussian*, where

$$\sigma^2 = \lim_{N\to\infty} \sum_{j=1}^{N} \frac{\langle|u_j|\rangle^2}{2} \tag{1.4}$$

From Equation 1.3, and taking into account that the intensity I and phase Φ of the resultant field are related to the real and imaginary parts of the field according to

$$U^{(r)} = \sqrt{I} \cos \Phi$$

$$U^{(i)} = \sqrt{I} \sin \Phi \tag{1.5}$$

it follows that the probability density of the intensity $p(I)$ and of the phase $P(\Phi)$ are given by

$$p(I) = \frac{1}{\langle I \rangle} e^{-\frac{I}{\langle I \rangle}} \quad \textit{for} \ \ I \geq 0 \tag{1.6}$$

and

$$p(\Phi) = \frac{1}{2\pi} \quad \textit{for} \ \ -\pi \leq \Phi \leq \pi \tag{1.7}$$

respectively. In Equation 1.6, $\langle I \rangle$ stands for the mean value of the intensity in the speckle diagram. Thus, according to the last two equations, the intensity distribution follows a negative exponential law, whereas the phase is uniformly distributed in the interval $(-\pi, \pi)$.

The moments of intensity distribution are defined as

$$\langle I^n \rangle = n!(2\sigma^2)^n = n!\langle I \rangle^n \tag{1.8}$$

and of special interest are the second-order moment and the variance:

$$\langle I^2 \rangle = 2\langle I \rangle^2 \ \text{and} \ \sigma_I^2 = \langle I^2 \rangle - \langle I \rangle^2 = \langle I \rangle^2 \tag{1.9}$$

This equation shows that the standard deviation of a polarized speckle pattern equals the mean value of the intensity. A usual measure of the degree of modulation of a speckle pattern is called *the contrast*, defined as

$$C = \frac{\sigma_I}{\langle I \rangle} \tag{1.10}$$

This definition, together with the result in Equation 1.9, means that the contrast of a polarized speckle pattern is always unity, and the speckle pattern is said to be fully developed.

1.3.2 SPATIAL EXTENT OF SPECKLES AND SPECKLE DIAGRAMS

In the preceding section, we have analyzed the statistical properties of a single point in space, or the first-order statistics of a speckle diagram, to describe the brightness fluctuations within the diagram. We are now going to devote some attention to the facts that determine the typical dimensions of the individual speckle grains. This detailed spatial structure is described by the second-order statistics. We are also going to discuss the system properties controlling the spatial extension of a diagram. It will be shown that the first is determined by the pupil of the system, whereas the angular extension of the whole diagram is mostly related to the microscopic details of the rough surface.

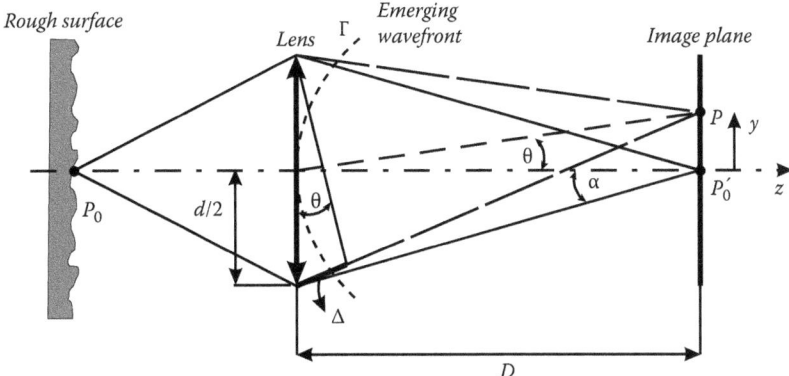

FIGURE 1.5 The lateral dimension of a speckle grain. The point P_0 is imaged by a lens of diameter d onto P'_0 at a distance D, where all wavelets from the wave front Σ are in phase. At another point, P, at the same image plane not all wavelets are in phase. The angle θ represents the height of P above optical axis and α is a measure of the aperture of the imaging lens.

A thorough and rigorous mathematical description of the second-order statistics of speckle diagrams involving autocorrelation calculations can be found in the classical book by Dainty.[4] Also, the work of Eliasson and Mottier,[35] and the paper from Goldfisher,[36] completely describe the autocorrelation of speckle patterns.

1.3.3 Typical Dimensions of Speckle Grains

Consider the schema shown in Figure 1.5 in which a point source P_0 is imaged by a lens, assumed to be perfect. We are not making any particular assumption about P_0, thus it can be taken as any individual point on the surface of the diffuser. The geometric image of P_0 is located at P'_0; the center of curvature of Σ, the emerging spherical wave, and thus, all the wavelets emerging from this wave front are in phase at P'_0. If we now consider another point P located at the same plane as P'_0 but laterally displacing a small distance y, the wavelets arriving at P will present some phase difference. The maximum phase difference will occur between points located at opposite extremes of the aperture, and the corresponding path difference can be easily evaluated to be

$$\Delta = d \sin \theta \qquad (1.11)$$

d being the lens diameter, and θ the angle defined by the optical axis and the line going from the center of the lens to the point P.

If the structure of the diffraction figure at P does not differ significantly from that at P'_0, it is necessary that the path difference remain much smaller than the wavelength λ. In symbols,

$$d \sin \theta \lll \lambda \qquad (1.12)$$

Or, because y is very small in comparison to the image distance D,

$$d \frac{y}{D} \lll \lambda \tag{1.13}$$

Equation 1.13 can be now used to define the lateral extension δ or diameter of the bright spot at P'_0, which can be taken as the "typical" size for a single speckle grain. Thus,

$$\delta = 2y \approx 2\lambda \frac{D}{d} = \frac{\lambda}{\alpha} \tag{1.14}$$

Being $2\alpha = \frac{d}{D}$ the aperture of the image-forming objective. Clearly, such a result is consistent, apart from having a constant factor for the circular aperture, with diffraction considerations.

To determine the axial extension of the bright spot, let us now consider a similar situation depicted in Figure 1.6. Following Françon,[37] let us assume that we axially displace the observation plane from π', where the image P'_0 of the source P_0 is located by an amount δz to a parallel position π''. It is also assumed that δz is small in comparison to the distance from the lens to π', shown as $D = OP'_0$ in Figure 1.6. Again, P'_0 being the image of P'_0, it is located at the center of curvature of the wave front Σ, and the waves emerging from different points of it are all in phase at P'_0. However, they present some phase difference at plane π'', which will give rise to a certain image deterioration. The path difference at P''_0 located on the axis at plane π'' can reach, at most, the value

$$\Delta = IP''_0 - OP''_0 \tag{1.15}$$

where I is the point at which the wave front Γ intercepts the ray emerging from the lens border. Note also that $IP'_0 = OP'_0 = D$.

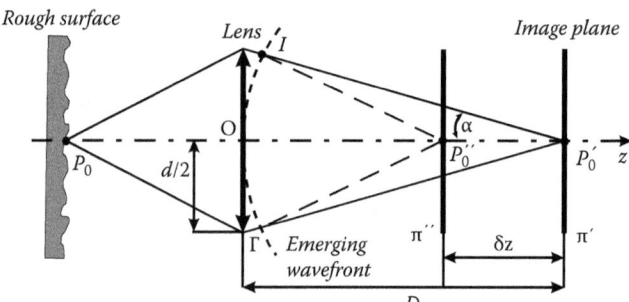

FIGURE 1.6 Geometrical considerations to determine the axial extension of speckles. The situation is similar to that shown in Figure 1.5, but now the image point P'_0 is compared to another point P''_0 axially displaced by a small step δz. Again, the angle α is a measure of the aperture of the imaging lens.

In accordance with the cosine theorem, and with reference to Figure 1.6, we can write

$$(IP_0'')^2 = D^2 + \delta z^2 - 2D\delta z \cos\alpha \qquad (1.16)$$

and also

$$OP_0'' = D - \delta z \qquad (1.17)$$

Equation 1.16 can be also written as

$$IP_0'' = D\sqrt{1 + \frac{\delta z^2}{D^2} - 2\frac{\delta z}{D}\cos\alpha} \approx D\left(1 - \frac{\delta z}{D}\cos\alpha\right) = D - \delta z\cos\alpha \qquad (1.18)$$

where it is assumed that $D \gg \delta z$ has been used. It follows now from Equations 1.15, 1.17, and 1.18 that

$$\Delta = D - \delta z\cos\alpha - D + \delta z = \delta z(1 - \cos\alpha) = 2\delta z \sin^2\frac{\alpha}{2} \qquad (1.19)$$

Finally, considering that the aperture α of the system is small, we can write for the maximum path difference:

$$\Delta = \delta z \frac{\alpha^2}{2} \qquad (1.20)$$

Using the preceding analogous arguments, we conclude that, if this path difference is much smaller than the wavelength, the image deterioration will be insignificant. This condition means that

$$\delta z \ll 2\frac{\lambda}{\alpha^2} \qquad (1.21)$$

and thus,

$$\delta z \approx 4\frac{\lambda}{\alpha^2} \qquad (1.22)$$

can be taken as a measure of the axial dimension of the bright spot.

These considerations are based on coherence arguments, and the conclusions are summarized in the results of Equations 1.14 and 1.22. These two equations give an order of magnitude for the dimensions of single speckle grains. Equivalent arguments[37] lead to the same results in the case of free space propagation. Speckle grains happen to be ellipsoids of revolution (cigar-shaped) of typical dimensions δ and δz, as shown schematically in Figure 1.7.

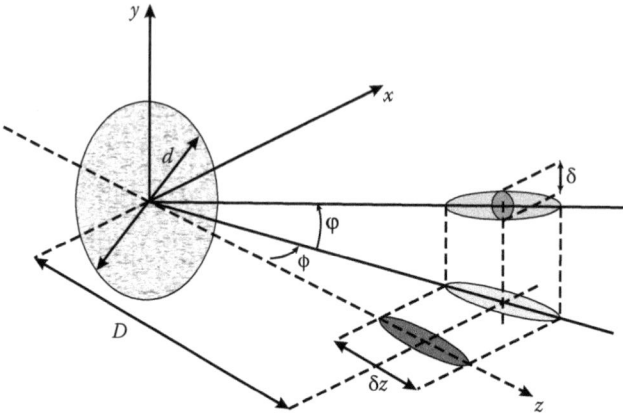

FIGURE 1.7 A perspective sketch showing the "cigar" structure of speckle grains in space with lateral and axial extents given by δ and δz, respectively.

Thus, independent of the type of optical setup producing a speckle pattern, image formations, or free space, the typical dimensions of the speckles are determined by the wavelength and the aperture of the system, diffraction being the optical phenomenon behind this.

Additionally, when the light source is partially coherent, a somewhat different speckle pattern is obtained.[11,14–16] In this situation, as in the coherent case, the average speckle size is determined by the autocorrelation function; but this function is the convolution of the coherent autocorrelation function with the light source autocorrelation function. Therefore, the average speckle size is larger than in the former case.[16] With respect to the speckle contrast, it is obviously less than or equal to 1.

1.3.4 THE SPECKLE HALO

We have already examined how a given optical configuration determines the typical lateral and axial extensions of single speckle grains produced by an optically rough object. We will now analyze the effect of the detailed structure of the rough surface upon the resulting illuminated field. Again, there are two totally equivalent approaches, namely, a reflecting rough surface, and a translucent glass with a grounded face. The latter situation is shown in Figure 1.8, where a collimated coherent beam illuminates a diffuser from its back. As in the earlier discussion, the diffuser is also limited by a stop of diameter d. For simplicity, the back surface of the glass is taken as flat, and the roughness is assumed to be localized at the exit surface of the object. Such height fluctuations of the uneven surface can be considered as a set of positive and negative lenses having randomly distributed focal lengths. Figure 1.8 shows two of those lenses, one convex, focusing the beam at point F_i, and the other, concave, from which light diverges as coming from focal point F_j. Applying similar reasoning to every portion of the diffuser (small, but large if compared to the illuminating wavelength), it can be replaced by a set $\{F_k\}$ of randomly distributed points acting as secondary emitters of coherent spherical wavelets of also randomly

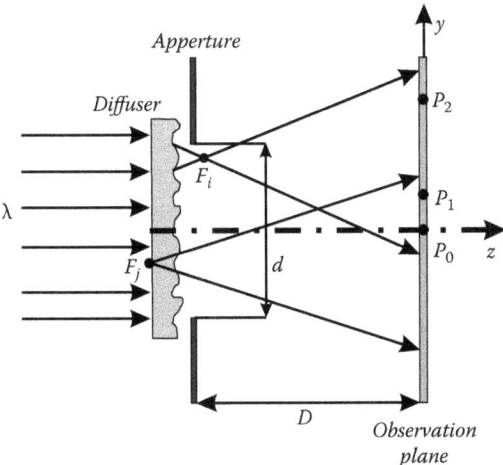

FIGURE 1.8 Origin of the speckle halo. A diffuser limited by a stop of diameter d is illuminated by a coherent field from its back. The total light intensity received by different points at the observation plane depends, in general, on the detailed microscopic structure of the rough surface.

distributed curvatures. Uniform distribution of phases is also warranted, provided the glass thickness presents random variations. As already discussed, this gives rise to a speckle pattern in the space in front of the diffuser. Consider now a couple of points P_1 and P_2 at a given observation plane, located at a distance D from the rough surface. P_1 is closer to the optical axis than P_2 and, as a consequence, it receives the contribution of amplitudes emerging from almost all points in the diffuser, whereas P_2 is only illuminated by those waves that are scattered at larger angles. As a result, the mean value of the intensity at the central portion of the observation plane will be, in general, greater than that at the periphery. This defines the illumination halo, and its actual profile will depend on the detailed structure of the diffuser (or the rough surface), such as the standard deviation of the height and curvatures of the individual scattering elements. For example, if all of them were identical, all emerging spherical wavelets would have the same angular spread. Thus, the mean intensity of the envelope of the pattern would be almost constant over a region at the observation plane, which corresponds approximately to the effective size of the diffuser given by the diameter d of the stop, and it would fall in a rather abrupt way outside this region. On the contrary, if the scattering elements have sizes that present a smooth distribution, the halo mean intensity will present a maximum at the center and will gradually decay as the distance to the center point, P_0, increases.

1.4 SPECKLE AS A RANDOM CARRIER OF INFORMATION

At the beginning of the 1960s, shortly after the first successfully laser operation, holography was a very promising field in optics, to which scientists devoted their main efforts. In those days, the speckle, always present when highly coherent optical fields are used, was seen mainly as a noise source, degrading the quality of holograms. However, the observation that the speckle was sensitive to the movement of the surface originating it led very quickly to the development of several techniques using its random intensity distribution as a carrier of information related to local changes in the surface, setting the basis for a new research field in coherent optics named *speckle metrology*. Very soon, several approaches were developed, including surface roughness measurements,[38] study of surface displacements and vibrations,[39] and image processing,[40] to mention a few. More recently, speckle techniques have been used in astronomy to go beyond the resolution limit set by atmosphere turbulence and achieve almost the diffraction limit of telescopes.[4]

Starting with the double exposure of a given pattern upon the same registering medium (photographic plate or CCD array), notable interference effects can be observed, constituting the base of the metrological applications of speckle. This was first noticed by J.M. Burch and J.M.J Tokarsky,[41] who proposed the multiple exposures of speckle patterns as a way to produce interference fringes, illustrating more the general principles than a given specific application. In their proposal, a given speckle diagram $D(\eta,\xi)$ is registered upon a high-resolution photographic plate H, η and ξ being the coordinates at the plane of H. After this, the plate is translated in the plane to a given distance ξ_0 parallel to the coordinate ξ, and the diagram is exposed again. The total intensity recorded on the plate is proportional to

$$I_{Tot.} \propto D(\eta,\xi) + D(\eta,\xi-\xi_0) \tag{1.23}$$

which can be written as the convolution of the speckle diagram with two Dirac delta functions displacing a distance ξ_0 with respect to each other:

$$I_{Tot.} \propto D(\eta,\xi) \otimes \left[\delta(\eta,\xi) + \delta(\eta,\xi-\xi_0) \right] \tag{1.24}$$

After development of the photographic plate, it is illuminated with a plane wave, and the transmitted light is observed at the focal plane of a lens. The resulting observed light intensity distribution at the focal plane of the lens is proportional to the Fourier transform of Equation 1.24, and consists of the spectrum $\mathcal{D}(u,v)$ of the speckle pattern modulated by a sinusoidal function arising from the transformation of the two Dirac delta functions, that is,

$$I(u,v) \propto |D|^2 \cos^2 \left[\frac{kv\xi_0}{2} \right] \tag{1.25}$$

where v is the coordinate in the Fourier plane associated with ξ, and $k = \frac{2\pi}{\lambda}$ is the wave number of the light of wavelength λ illuminating the plate.

The spatial frequency of the fringes modulating the spectrum of the speckle pattern, which is also of random characteristics, codifies the information about the translation of the plate. In the experiment described earlier, which represents the technique known as *speckle photography*, the speckle pattern remains the same between both exposures; this implies that both the recorded patterns are fully cor-related over the entire area of the register medium, thus maximizing the visibility of the resulting fringes.

Another approach in the field of speckle metrology, known as *speckle inter-ferometry,* was first proposed by Leendertz[39] at the beginning of the 1970s; in it the desired information is codified in the correlation fluctuations of the involved speckle patterns. This technique, together with the fast development of computers and image-processing systems, such as digital cameras and frame grabbers capable of many built-in operations, allowed the appearance of the first *optodigital* devices, leading to a technique known today as *electronic speckle pattern interferometry* (ESPI) or, alternatively, as *TV holography.* ESPI was first proposed at Loughborough University, England[42] and simultaneously in Standford in the United States[43] as an attempt to simplify the rather cumbersome and slow procedures of speckle photog-raphy involving chemicals to develop the plates and films. Since the late 1980s, ESPI has been used as an invaluable tool in research and industry to carry on nondestruc-tive testing.[44]

Because of its relevance in speckle metrology, we will devote a few paragraphs to better describe this technique, albeit far from making a very detailed analysis of it. In the early years of the 1970s, scientists recognized the possibility of applying the basic ideas of holographic interferometry by using a video system for record-ing and processing signals, instead of the usual photographic methods. The main interest was to find and develop a new generation of quick systems capable of real time or almost real time operation, making them useful for applications such as online inspection of components during fabrication. The result was a novel tech-nique combining the interferometric sensitivity of holography and speckle with the versatility of electronic data processing. In the basic setup of an ESPI, shown in Figure 1.9, the rough object to be studied is illuminated by a properly expanded laser beam and imaged by a CCD camera. A second coherent beam, acting as a reference, is superimposed at the sensor plane onto the image of the object via the beam splitter BS2. This beam can be chosen so as to emerge from the front focal plane of the imaging system, thus providing a plane reference wave front at the camera sensor. The aperture of the system has to be small enough to ensure that a single speckle is larger than a pixel of the CCD sensor. Otherwise, several speckle grains could fall onto a single sensor pixel, degrading the contrast of the whole pattern. The resulting signal is displayed on a TV monitor. The crucial point here is that, before the signal reaches the monitor, it can be electronically manipulated to show a desired operation. In particular, after the given (reference) state of the object under study has been registered, it can be translated, deformed, or subjected to some stress before the new situation is recorded, thus producing the correspond-ing modifications in the speckle pattern. If the intensity diagram registered for the reference situation is subtracted pixel by pixel from the intensity corresponding to the new situation, the resulting difference image will present intensity variations in correlation with the corresponding phase changes (modulus 2π) between the two exposures.

In fact, for those points where phase has changed an integer multiple of 2π radi-ans between exposures, both speckle diagrams will be fully correlated and, as a

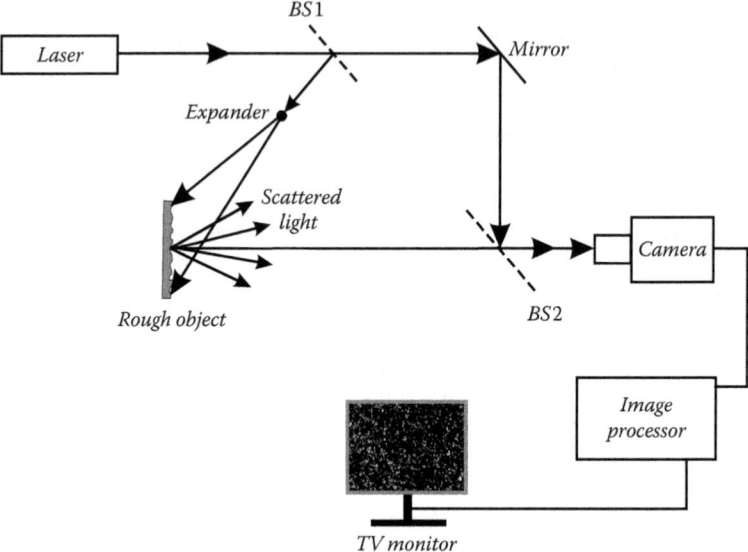

FIGURE 1.9 Schema of the experimental setup of an electronic speckle pattern interferometer.

consequence, the difference in their intensities will vanish, showing a dark spot. The other extreme is represented by zones where the phase has changed by an odd multiple of π radians between exposures. The intensities between both exposures present large changes, and the difference shows bright spots. In general, those points that had experienced arbitrary phase changes will appear, after subtraction, with intensities between these two extreme values. In summary, after subtraction, the object appears covered by a speckle diagram modulated by correlation fringes following zones of equal phase changes. As said, the correlation fringes represent the phase change modulus 2π and, because of this, the phase needs to be unwrapped before useful information can be retrieved.

These ideas can be readily formalized as shown in the following text. Considering again the sketch of Figure 1.9, a laser beam is divided into two beams by the BS1. The object to be studied is illuminated by one of the beams and imaged by the CCD camera, whereas the coherent beam is directed to the camera by using another beam splitter. As a result, at a given point on the CCD sensor, the total field is the sum of the complex amplitudes of the object and reference fields, given by

$$U_0(r) = u_0(r)e^{i\phi_0(r)} \tag{1.26}$$

and

$$U_R(r) = u_R(r)e^{i\phi_R(r)} \tag{1.27}$$

respectively. For an optically rough surface, u_0 and ϕ_0 vary in a random way, giving rise to a speckle diagram represented by

$$I_1(r) = U_0 U_0^* + U_R U_R^* + U_0 U_R^* + U_R U_0^* \qquad (1.28)$$

In the right member of Equation 1.28, the explicit dependence with r has been omitted to simplify the notation. The image corresponding to I_1 is now stored, and then some phase change $\varphi(r)$ is introduced at r by translating the object and/or by means of some deformation of it. The new complex amplitude of the object field is now given by

$$U_0'(r) = u_0(r) e^{i[\phi_0(r) + \varphi(r)]} = U_0(r) e^{i\varphi(r)} \qquad (1.29)$$

In Equation 1.29, it has been assumed that changes are small enough to neglect alterations in the scattered amplitude. The intensity registered by the camera is now given by

$$I_2(r) = U_0 U_0^* + U_R U_R^* + U_0 U_R^* e^{i\varphi(r)} + U_R U_0^* e^{-i\varphi(r)} \qquad (1.30)$$

Subtraction of Equation 1.28 from Equation 1.30 gives the resulting intensity:

$$I(r) = I_2(r) - I_1(r) = U_0 U_R^* (1 - e^{i\varphi(r)}) + U_R U_0^* (1 - e^{-i\varphi(r)}) \qquad (1.31)$$

Equation 1.31 clearly shows that $I(r)$ will be zero at those points for which the introduced phase change $\varphi(r)$ is zero or an integer multiple of 2π, whereas for other arbitrary phase changes, $I(r)$ will present a sinusoidal variation taking values between zero and the maximum value that can be represented in the image processor, giving rise to correlation fringes. Those zones where $\varphi(r) = 2(n+1)\pi$ can be taken as the geometrical places at which the fringes present maximum intensity. Notice that negative intensity values cannot be shown on a video monitor. Equation 1.31 is thus displayed in modulus on the TV screen.

1.5 DYNAMIC SPECKLE

When laser light giving rise to the speckle pattern is scattered by objects showing some type of activity (fruit, paints in process of drying, or some kind of biological samples), the visual appearance of the pattern is similar to that of a boiling liquid. Such an effect is called *dynamic speckle,* and it is caused by variations in the phase of light produced by movements of the scattering centers,[45,46] changes in the refractive index, rotary power, etc. Light coming from different scatterers produce low frequency beats in the detector, which appear as variations in local intensity. Several papers using such techniques report the measurement of the dynamic activity of seeds,[47] bruising in fruit,[48] parasite activity,[49] etc. As the detailed description of this phenomenon will be the central point of the following chapter, we will stop now.

REFERENCES

1. Goodman, J. W., Statistical properties of laser speckle patterns, *Technical Report No. 2303-1*, Stanford Electronics Laboratories, Stanford University, 1963.
2. Goodman, J. W., Some fundamental properties of speckle, *J. Opt. Soc. Am.,* 66, 1145, 1976.
3. Goodman, J. W., Statistical properties of laser speckle patterns, in *Laser Speckle and Related Phenomena*, 2nd ed., Dainty, J. C., Ed., Springer Verlag, New York, 1984, chap. 2.
4. Dainty, J. C., *Laser Speckle and Related Phenomena*, 2nd ed., Dainty, J. C., Ed., Springer Verlag, New York, 1984.
5. Dainty, J. C., The statistics of speckle patterns, in *Progress in Optics*, Wolf, E., Ed., North-Holland, 1976, 14.
6. Mas, G., Palpacuer, M., and Roig, J., Spectre de granularité d'une plage diffusante éclairé en lumière cohérente. Calcul du contraste, *C. R. Acad. Sc. Paris t.*, 269B, 633, 1969.
7. Allen, L. and Jones, D. G. C., An analysis of the granularity of scattered optical maser light, *Phys. Lett.* 7, 321, 1963.
8. Considine, P.S., Angular dependence of radiance of rough surfaces in imaging systems, *J. Opt. Soc. Am.,* 56, 877, 1966.
9. Isenor, R., Object-image relationships in scattered laser light, *Appl. Opt.*, 6, 163, 1967.
10. Enloe, L. H., Noise-like structure in the image of diffusely reflecting objects in coherent illumination, *Bell Sys. Tech. J.*, 46, 1479, 1967.
11. Martienssen, W. and Spiller, E., Coherence and fluctuations in light beams, *Am. J. Phys.*, 32, 919, 1964.
12. Mas, M., Roig, J., and Delour, J. P., Granularité d'une plage diffusante éclairé en lumière cohérente. Observations dans le plan focal, *C. R. Acad. Sc. Paris t.* 269B, 55, 1969.
13. Mas, G., Thèse d'Etat, Docteur ès Sciences Physiques, Université de Montpellier, 1969.
14. Dainty, J. C., Some statistical properties of random speckle patterns in coherent and partially coherent illumination, *Optica Acta*, 17, 761, 1970.
15. Parry, G., Speckle patterns in partially coherent light, in *Laser Speckle and Related Phenomena*, 2nd ed., Dainty, J. C., Ed., Springer Verlag, New York, 1984, chap. 3.
16. Gaggioli, N. G., *C.R. Acad. Sc. Paris t.*, 275B, 727, 1972.
17. Sporton, T. M., The scattering of coherent light from a rough surface, *Br. J. Appl. Phys. (J. Phys. D)* 2, 1027, 1969.
18. Rigden, J. D. and Gordon, E. I., The granularity of scattered optical maser light, *Proc. IRE*, 50, 2367, 1962.
19. Oliver, B. M., Sparkling spots and random diffraction, *Proc. Inst. Elect. Electron. Engrs.*, 51, 220, 1963.
20. Langmuir, R. V., Scattering of laser light, *Appl. Phys. Lett.*, 2, 29, 1963.
21. Raman, Sir C. V. and Datta, G. L., On Quételet's rings and other allied phenomena, *Phil. Mag.*, 42, 826, 1921.
22. Newton, Sir I., *Opticks*, Book I, Part I, Prop VIII, Prob. II (Reprinted by Dover Press, New York, 1952, pp. 107–1730.
23. Rayleigh, Lord, Note on the explanation of coronas, as given in Verdet's *Lecons d'Optique Physique* and other works, *Proc. Lond. Math. Soc.*, 3, 267, 1871.
24. Exner, K., Über die Fraunhofer'schen Ringe, die Quetelet'schen Streifen und verwandte Erscheinungen, *Sitzungsber. Kaiserl. Akad. Wiss. (Wien)* 76, 522, 1877.

25. Laue, M. von, Die Beugungserscheinungen an vielen unregelmäßig verteilten Teilchen, Sitzungsberichte der koniglich preussischen Akademie der Wissenscxhaften (Berlin) 44, pp. 1144–1163, 1914.
26. de Haas, W. J., On the diffraction phenomenon caused by a great number of irregularly distributed apertures or opaque particles, *KNAW Proceedings*, 20 II, 1278, 1918.
27. Schiffner, G., Die Granulation im diffus gestreuten Laser Licht, Dissertation, Vienna, 1966.
28. Schiffner, G., Granularity in the angular spectrum of scattered laser light, *Proc. IEEE* 53, 1245, 1965.
29. Ratcliffe, J. A., Some aspects of diffraction theory and their application to the ionosphere, *Rep. Progr. Opt.*, XIX, 188, 1956.
30. Allen, L. and Jones, D. G. C., An analysis of the granularity of scattered optical laser light, *Phys. Lett.*, 7, 321, 1963.
31. Hariharan, P., Speckle patterns: A historical retrospect, *J. Mod. Opt.*, 19, 791, 1972.
32. Raman, C. V., The scattering of light in the refractive media of the eye, *Phil. Mag.*, 38, 568, 1919.
33. Buchwald, E., *Ber. Dt. Phys. Ges.*, 21, 492, 1919.
34. Ramachandran, G. N., Fluctuations of light intensity in coronae formed by diffraction, *Proc. Indian Acad. Sci.*, A2, 190, 1943.
35. Eliasson, B. and Mottier, F., Determination of the granular radiance distribution of a diffuser and its use for vibration analysis, *J. Opt. Soc. Am.*, 61, 559, 1971.
36. Goldfischer, L. I., Autocorrelation function and power spectral density of laser-produced speckle pattern, *J. Opt. Soc. Am.*, 55, 247, 1965.
37. Françon, M., *Laser Speckle and Applications in Optics*, Academic Press, New York, 1979.
38. Lèger, D. and Perrin, J. C., Optical surface roughness determination using speckle correlation technique, *Appl. Opt.*, 14, 872, 1975.
39. Leendertz, J. A., Interferometric displacement measurement on scattering surfaces utilizing speckle effect, *J. Phys. Eng. (Sci. Inst.)*, 3, 214, 1970.
40. Debrus, S. et al., Interference and diffraction phenomena produced by a new and very simple method, *Appl. Opt.*, 8, 1157, 1969.
41. Burch, J. M. and Tokarsky, J. M. J., Production of multiple beam fringes from photographic scatters, *Optica Acta*, 15, 101, 1968.
42. Butters, J. N. and Leendertz, J. A., Holographic and video techniques applied to engineering measurement, *Trans. Inst. Meas. Ctrl.*, 4, 349, 1971.
43. Macovski, A., Ramsey, S. D. and Schaefer, L. F., Time-lapse interferometry and contouring using television systems, *Appl. Opt.*, 10, 2722, 1971.
44. Stroh, R. S., Ed., *Speckle Metrology*, Marcel Dekker, New York, 1993.
45. Briers, J. D., The statistics of fluctuating speckle patterns produced by a mixture of moving and stationary scatterers, *Opt. Quantum Electron.*, 10, 364, 1978.
46. Briers, J. D., Laser doppler and time-varying speckle: A reconciliation, *J. Opt. Soc. Am.*, A 13, 345, 1996.
47. Arizaga, R. et al., Display of local activity using dynamic speckle patterns, *Opt. Eng.*, 41, 287, 2002.
48. Pajuelo, M. et al., Bio-speckle assessment of bruising in fruits, *Opt. Lasers Eng.*, 40, 13, 2003.
49. Pomarico, J. A. et al., Speckle interferometry applied to pharmacodynamics studies: Evaluation of parasite motility, *Europ. Biophys. J.*, 33(8), 694, 2004.

2 Dynamic Speckle
Origin and Features

Marcelo Trivi

CONTENTS

2.1 INTRODUCTION

When a coherent beam coming from a laser illuminates a rough object, a typical granular interference pattern named speckle is observed.[1-3] Besides, laser light scattered from diffuse objects produces a similar pattern.

If the surface of the objects does not remain rigid but rather presents some type of local movement, then the intensity and the shape of the observed speckle evolves in time. The speckle patterns thus become time dependent. This phenomenon is characteristic of biological samples[4,5] and it is named biospeckle. The visual appearance of the speckle diagram is similar to that of a boiling liquid.

The study of the speckle patterns, due to their random nature, requires using statistical tools. The statistical properties depend on the coherence of the incident light and the characteristics of the diffusing surface.

The dynamic statistical properties of speckles have been extensively studied theoretically and experimentally for vibrations,[6] velocity measurements,[7] and displacements.[8] These studies are presented as a simple scattering at the surface of rigid and inanimate objects. In the case of living samples, dynamic speckles have different characteristics, and multiple scattering must be considered. Also, speckle patterns are formed by different proportions of moving, scatters with different dynamics, including static speckles. The space–time speckle dynamics depends on the structure and activity of the living samples. Therefore, theoretical description is difficult, and numerical approaches have been developed.[9-11]

The observation of biospeckles indicates that they generally fluctuate in a space–time fashion. This is due to the complicated structure and activity of living objects. Thus, the dynamic behavior of biospeckles should be analyzed statistically. Although the accurate mathematical description of different dynamic speckle patterns is very difficult, a statistical analysis can be established. Some statistical approaches have been taken for the intensity fluctuation of biospeckles, which will be developed in Chapter 4.

Usually, there are two types of dynamic speckles:[12] *translational speckles* and *boiling speckles*. The first ones are those whose shapes remain unchanged even after considerable displacements when a diffuser moves and the entire speckle grains move as a whole. This phenomenon occurs in solid diffuse object displacements, and the speckles are called translational speckles. The second ones are those that, when a diffuser moves, deform, disappear, and reappear without any significant displacement of their position, and are called boiling speckles. These are typical of living objects. Each type behaves considerably different from the other.

This behavior of the speckle pattern can also be observed in nonbiological industrial processes, including the drying of paint,[13] corrosion,[14] evolution of foams,[15] salt efflorescence on stone surfaces,[16] hydroadsorption in gels,[17] etc.

The dynamic activity takes place when the sample changes its properties due to movement of the scattering centers, changes in the optical path due to variations of the refractive index, configuration changes, combination of these situations, bubbles explosions, etc.

If a diffuse object moves, the individual grains of the speckle pattern also move and change shape; the moving speckle pattern thus contains information about the object in motion. Many efforts have been carried out to assign numbers that characterize this activity, and that correlate favorably with alternative measurement methods of interest for the experimenter. The study of the temporary evolution of speckle patterns may provide an interesting tool to characterize the parameters involved in these processes.

Biospeckles carry useful information about the biological and physiological activity of biological samples. With the extended use of laser technology in the medical and biologic fields, the applications of biospeckle techniques have been increasing in the last years, for example, measurements of blood flow,[18] parasite motility,[19] atherosclerotic plaque,[20] cerebral activity in rats,[21] characterization of regional mesenteric blood flow,[22] health tumor tissue,[23] botanical specimens,[24] etc.

In this chapter, we introduce some concepts to discuss the origin and features of the biospeckle phenomenon.

We start with a brief description of the various ways that the light interacts with matter, in particular, with the applications of lasers in biology and medicine based on the use of numerous phenomena associated with various effects of interaction of light with biological objects. Light can be reflected, scattered, absorbed, or reradiated when it interacts with bioobjects. These phenomena are analyzed using Fresnel coefficients, Lambert–Beer law, and the scattering theory. As the biospeckle phenomenon is characteristic of biological samples, we present a description of the cells that are the main components of living tissues, and we analyze their optical properties. A brief description of the statistical properties of biospeckle patterns, as such size, intensity, and depth, are included. Chapter 4 is devoted to presenting an in-depth analysis of these properties. We have also included some examples of the biospeckle phenomenon and its origin in nonbiological samples. We present cases of paint drying, corrosion, foams, and salt efflorescence. These cases will be commented in Chapter 9.

Speckles are modulated by the state of motion of the scatterers. The frequency of speckle fluctuations is related to the change in the random walk and the velocity of each scatterer and can be compared with the Doppler effect and light-beating spectroscopy.[26,27] Therefore, we present a biospeckle description as a Doppler effect using an holodiagram[28] and sensitivity vector scheme. We have also included an analysis by Briers,[29] where he demonstrates the equivalence among time-varying speckle phenomenon, laser Doppler velocimetry, and photon correlation spectroscopy.

Other aspects that are also analyzed are the effects produced by changes in the refractive index of the samples or the configurational changes in the experimental setup, light source, or detectors. The biospeckle pattern also depends on the phase changes of many different waves of light that interfere to produce it. So, we discuss other phenomena, such as turbulence and diffusion that introduce biospeckle fluctuations.

2.2 LIGHT–MATTER INTERACTION

Light interacts with matter in several different ways. The processes that occur depend on the wavelength of the light and the structure of the medium. Light can be reflected, scattered, absorbed, or reradiated when it interacts with matter. These

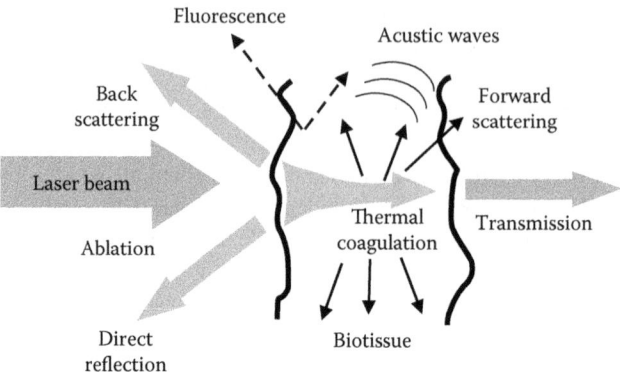

FIGURE 2.1 Different interactions between light with biologic tissues.

interactions could be occurring several times. In particular, the application of lasers in biology and medicine is based on the use of numerous phenomena associated with various effects of interaction of light with biological objects. In this section, we follow the discussion of V. V. Tuchin.[30] Figure 2.1 shows different interactions between light and biologic tissues.

Different processes of laser radiation interaction with biological objects can take place: nonperturbing processes for those not having a pronounced effect on the samples, processes with photochemical or thermal action, and processes that cause photodestruction. Photons with very high energy, such as gamma rays and x-rays, may even ionize atoms or break bonds in the molecules; but this will not be the case in this book because low-intensity visible light is being used in all the cases, and we have considered only the first group processes.

The influence of light on the functioning of living matter depends on the degree of homeostasis of the bioobject and the evolution level. It is the lowest for biological molecules, and the highest for vertebrates.[30] Homeostasis is the property of either an open system or a closed system, especially a living organism that regulates its internal environment so as to maintain a stable, constant condition. Multiple dynamic equilibrium adjustments and regulation mechanisms make homeostasis possible. Low-intensity radiation does not activate the adaptation mechanisms of a biosystem and does not affect its homeostasis. Thus, at small intensities, it is possible to study the processes occurring in living objects without any serious perturbation in their behavior. In our case we can employ biospeckle techniques, which do not need strong light and do not distort the measurement results because the homeostasis of living matter remains intact even at the local level.

2.2.1 The Optics of Bioobjects

Biological objects are optically inhomogeneous absorbing media, and their refractive index is higher than that of air, which is typically close to the refractive index of water. When a laser beam is incident on the air/bioobject interface, part of the radiation is reflected and the other part penetrates into the object. There, absorption and multiple scattering occur, and the laser beam broadens and weakens. Besides, volume scattering causes backpropagation.

The reflection and transmission of light, when it enters a border between different refractive indexes, obey the elemental Snell's law (Equation 2.1) and the Fresnel coefficient (Equation 2.2). It depends, therefore, on the refractive indexes as well as the angle of the incoming and the reflected rays:

$$n_1 \cdot \sin \varphi_1 = n_2 \cdot \sin \varphi_2 \tag{2.1}$$

$$R = \frac{1}{2} \left[\frac{\sin^2(\varphi_1 - \varphi_2)}{\sin^2(\varphi_1 + \varphi_2)} + \frac{\tan^2(\varphi_1 - \varphi_2)}{\tan^2(\varphi_1 + \varphi_2)} \right] \tag{2.2}$$

where n_1, n_2 are the refractive indexes in the two materials, and φ_1, φ_2, the angles of the light with the perpendicular to the boundary; R is the reflectance coefficient.[31]

2.2.2 SCATTERING

Bioobjects are characterized by considerable light scattering.[32] In general, it consists of many scattering centers randomly distributed over their volume. An example of an exception is the transparent tissues of the eyes.

Scattering can be either elastic or inelastic. Elastic means that the scattered photons neither lose nor gain energy in the process. The types of elastic scattering are *Rayleigh* and *Mie*. In Rayleigh scattering, the electromagnetic field of the incoming light induces a polarization of the molecule, which "reradiates" the light with retained wavelength and without delay. This appears when the particles in the substance are about the same size as or are smaller than the wavelength of the light.

The scattering cross section σ_{scat} relates the scattering of light or other radiation to the particles present. Its SI unit is the square meter (m²), although smaller units are usually used in practice. The term *cross-section* is used because it has the dimensions of area. For Rayleigh scattering, the cross section is about 10^{-26} cm^{-2}, and the intensity of the scattering light is proportional to λ^{-4}, which means that it increases as the wavelength decreases.[33] For instance, the blue light from the sun is scattered on the molecules in the air more than the red light, and it therefore makes the sky look blue.

If the particles are bigger than the wavelength, there will be Mie scattering. In this case, in a simple model, a ray penetrates the wall of the particle and reflects once or several times against the inside of the wall before it leaves the particle. Examples of the Mie scattering are the bad visibility in fog, and the rainbow in the clouds. In the latter example, the sunlight scatters in the waterdrops in the clouds. As the sunlight consists of wide spectra, the drops in the clouds will operate as small prisms that scatter the different wavelengths in different angles to the sun, and there will be a rainbow. The cross section for Mie scattering depends very much on the sizes and refractive indexes of the particles and the surrounding media. It varies, therefore, between 10^{-26} cm² and 10^{-8} cm².[33] In this case, the intensity of the scattering light is proportional to λ^{-2}.

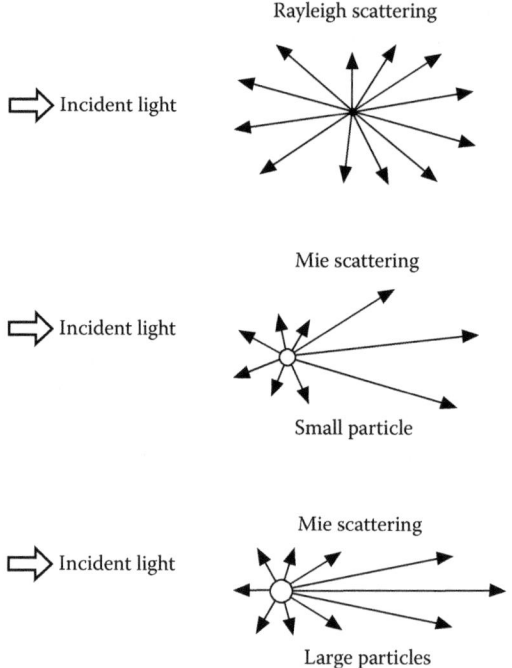

FIGURE 2.2 An example of Rayleigh and Mie scattering on particles of different sizes.

Figure 2.2 shows an example of Rayleigh and Mie scattering on particles of different sizes.

Related to inelastic scattering are the two strongest inelastic effects in scattered light: Raman and Brillouin effects. The first one is a special case of the Rayleigh scattering. Rayleigh and Raman scattering differ in the reradiation of light. In Raman scattering when the electrons are excited they fall back to a different energy level from the one they were before excitation. This gives the wavelengths certain energies, up or down, from the wavelength of the incoming light, specific for each bond in the medium.

Hydrogen has the largest shift, and changes the energy 4155 cm^{-1}, but the most common shifts are 100–1000 cm^{-1}.[34] The cross section for Raman scattering is about 10^{-29} cm^2. Brillouin scattering appears when a crystal is deformed by a long acoustic phonon.[35] The refractive index of the crystal changes because of the tension that is induced by the vibration. Consequently, if a phonon is present, it will affect the shifts and directions of all the photons it encounters. The size of the change depends on the resonance frequencies of the setup, and hence, the characteristics and size of the object, laboratory, equipment, and phonon.

2.2.3 ABSORPTION

Light absorption is one of the factors that characterize the interaction of light with biological tissues. In the ultraviolet (UV) and infrared (IR) spectral regions, absorption dominates, and the contribution of scattering is minor.[30] Therefore, the penetration of the light is small, sometimes only 50–100 μm. In the visible spectra region, both absorption and scattering are important. The penetration depth for a typical biotissue is 0.5–2.5 mm.

The absorption process obeys the Lambert–Beer law. The intensity I of light transmitted by a biological object is

$$I(z) = (1 - R) I_0 \exp(-kz) \tag{2.3}$$

where R is the Fresnel reflection coefficient, I_0 is the intensity of the incident light, $k = \alpha + s$ is the attenuation coefficient, α is the absorption coefficient, s is the coefficient of loss due to scattering, and z is the specimen probe depth.

Most biotissues have appreciable absorption coefficients. In these cases, $\alpha \gg s$, and therefore very thin specimens are needed to determine the value of α:

$$\alpha \approx k \approx \frac{1}{d} \ln \left[\frac{I_0'}{I(d)} \right] = \frac{1}{d} \ln \left(\frac{1}{T} \right) \tag{2.4}$$

where d is the specimen thickness and T is the transmission coefficient of the tissue.

$$I_0' = (1 - R) I_0 \tag{2.5}$$

For example, for soft tissues, $\alpha \sim 600$ cm^{-1} at $\lambda = 10.6$ μm, and hence, $d = 38$ μm at $T = 0.1$.

The absorption process is a very complex part of the interaction. The different atoms and molecules in matter have a wide range of possible energy levels that can be excited. Depending on the type of atoms, after a certain period of time they lose energy[33] either by producing heat, contributing to photochemical reactions, or reradiating photons in any direction (called *fluorescence*). Depending on the occupation of the different energy levels, the photons can be either of the same or different wavelengths compared to those of the incoming rays.

Each medium has specific characteristics; consequently, it is useful to obtain information about the substance when working with spectroscopy. However, because this study depends on the scattering correlation in different media, reradiation can be a source of noise. The probabilities for absorption and fluorescence are both normally about 10^{-16} cm^2. However, in liquids and solids at atmospheric pressure, the molecules will be close enough to collide into each other. This is called *quenching*, and it reduces the probability of fluorescence to about 10^{-20} cm^2.[33]

The most common way of describing the properties of a medium is with the absorption coefficient μ_a and scattering coefficient μ_s. To get the total coefficient, the two are coupled to define the linear transport coefficient:

$$\mu_{tr} = \mu_s' + \mu_a \tag{2.6}$$

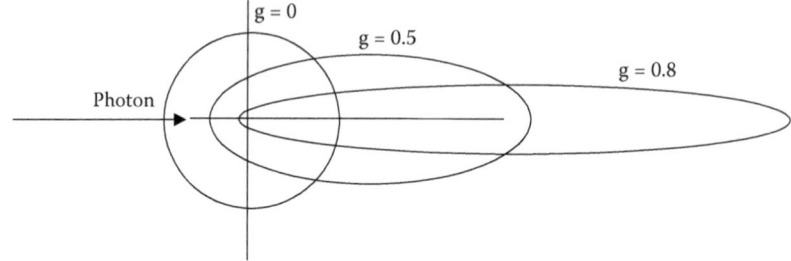

FIGURE 2.3 The probability for a photon to be scattered in certain directions for different values of *g*.

where $\mu \otimes_s$ is the reduced scattering coefficient.

$$\mu'_s = (1-g)\mu_s \qquad (2.7)$$

which depends on the anisotropy factor g, where it is the mean cosine of the scattering angle and can attain values between −1 and 1. When $g = 1$, it represents a pure forward scattering, and when $g = 0$, it means a uniform scattering (see Figure 2.3). Human tissue, for example, has a typical value between 0.7 and 0.95 for the factor g.[36]

2.3 BIOLOGICAL TISSUES: COMPOSITION AND STRUCTURE

The biospeckle phenomenon is characteristic of biological samples. So, in this section, we present a brief description of biological tissues.

The main components of living tissue are cells.[25] Biological studies show great diversity in size, morphology, and metabolism, and a complex structure and function for each component of the cell. They perform many important functions, for example, the synthesis of blood protein, carbohydrate storage, destruction of toxic substances, photosynthesis, reproduction, etc.

The cell is composed of different "organelles," such as the nucleus, mitochondria, chloroplasts, lysosomes, peroxisomes, cavities of the endoplasmic reticulum, Golgi apparatus, etc. A typical animal cell with its organelles is shown in Figure 2.4. Each organelle has a particular morphology, a different biochemical composition, and function. In addition, the organelle units are not static in size, shape, or function, and relations with other organelles can change. Every unit moves; some can grow, double, and renew. The simplest cells are not more complicated than the mitochondria. The more complicated cells have an intricate association representing thousands of organelles. Each cell type is characterized by a particular organization of its organelles and macromolecules. A complex cell as rat hepatocyte, which is cited in some books[25] as an example of an animal cell, contains between 1000 and 2000 mitochondria, about 500 peroxisomes, several hundred lysosomes, several million ribosomes, a few centrioles, and a very large number of microfilaments and microtubules.

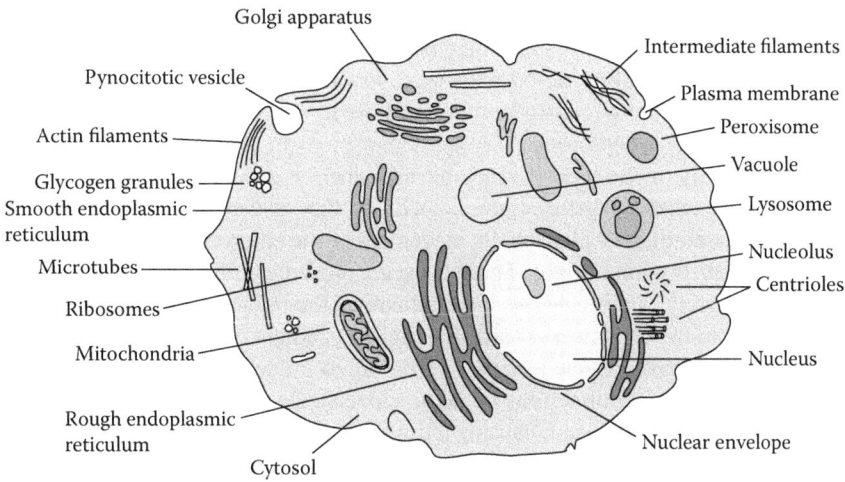

FIGURE 2.4 A typical animal cell with its organelles.

The differences in structure between cells are mainly due to differences in roles. For example, in plant cells chloroplasts exist, which are responsible for photosynthesis, but they are not present in animal cells.

The cells that share certain functions often have a similar organization. For example, almost all animal cells perform protein synthesis and then secrete these molecules through the cell membrane. They have large nucleoli and rough but highly developed endoplasmic reticulum.

Membranes divide the cells into different morphological and functional compartments. Also, membranes limit organelles within the cell. Membranes are not simple mechanical barriers but very complex molecular sets that consist mostly of lipids and proteins that contain enzymes. They regulate the quantity and nature of substances that can move from the cell to the environment and vice versa, as well as from one intercellular compartment to another.

The cell diameters range from 0.1 μm in the simplest procariotides to 1–100 μm for the protozoa, algae, and pluricellular organisms. However, some specialized cells, for example, egg cells, reach diameters on the order of millimeters and centimeters.

The cell wall is a 30-nm thick layer of polysaccharides and proteins that will stretch as the cell grows and eventually deteriorate and break as the cell dies. In the cell there are several nuclei with the size of about 0.5 μm, which includes the DNA. The mitochondria have about the same size as the nuclei and control the respiration of the cell, as it provokes the oxidation of nutrients. In vegetable cells, there are also larger parts, namely chloroplasts, which are the engines of the cells. They are a few micrometers in diameter, contain chlorophyll, and are responsible for the production of vegetable energy. The ligament that glues the particles together consists of endoplasmic reticulum, which produces and transports proteins to the membranes. The biggest part of the cell is the vacuole, which is a reservoir of water, sugars, pigments, oxygen, and carbon dioxide. It is the part that expands the most and makes the cell grow.

Cells stick to each other like bricks in a wall even though there is interspace between some of them. This space may be filled with air or loose subparts of cells. The surface is normally covered with wax, small fluff called *trichomes*, and pigments to moderate the wavelengths that penetrate it.

The distribution of chloroplasts is not homogenous in its way through the tissue. It depends highly on the type of vegetable and where it grows. When it grows in the shade and is reached by diffuse light, it needs another distribution than if it receives direct collimated light. It also entails movement of the chloroplasts, called *cyclosis*, to adjust the absorption of light. There is even a movement of whole cells to use the light as effectively as possible. As light penetrates the tissue, its intensity decreases with depth, and the cells turn towards the highest gradient of the light.

Thus, the biological tissue is actively working to influence light propagation and absorption. Therefore, this activity could be correlated with fluctuations in the intensity of the light at the detector, causing a time-varying speckle pattern.

2.3.1 OPTICAL PROPERTIES OF BIOLOGICAL TISSUE

Any transparent matter with a curved surface and a refractive index higher than air can focus light. The tissue has absorption centers consisting of so-called chromophores. It is in these centers that the tissue can optimize its use of light; consequently, cells concentrate the incoming light on these areas to store energy. The electromagnetic field in biological tissue is very heterogeneous and the concentrations of light vary a lot, and therefore multiple scattered light must be considered.

As mentioned earlier, reflection, scattering, and absorption are different types of interactions of light with matter. In order to think microscopically about scattering in connection with these interactions, it can be said that, because the tissue has a higher refractive index than air, there will be a reflection on the surface. However, the penetrating light can be also trapped inside. If the light used is polarized, it will lose its polarization at a high exponential rate when penetrating the tissue. Therefore, almost all the light reflected back with the kept polarization comes from surface reflection.[37] When illuminating biological tissue with polarized light, the polarization in the specular direction of observation was 0.98 due to reflection in the surface but only 0.17 in a 0°/30° setup.

The refractive indexes are different inside and outside the cells as well as in all the cell parts and in the cell wall. The indexes of different cell walls have been investigated, with results between 1.333–1.472 and an average of 1.425.[38] This means that the backscattered light from biological tissue is not only multiple scattered but also multiple reflected on different surfaces.

Biological tissue is a highly scattering media. The sizes of the cells, and the vacuoles and spaces between the cells, are a bit larger than the wavelength of a He–Ne laser. Thus, they are probably participating in Mie scattering, which consequently is very strong, with a cross section comparable with the absorption[33] 10^{-16} cm^2. All the other parts as well as all the molecules are participating in the Rayleigh scattering. The cross section for Rayleigh scattering, 10^{-26} cm^2, is however small compared to that for Mie scattering. This also leads to the Raman effect, which is very small, as the cross section is about 10^{-29} cm^2.

There is also Brillouin scattering in biological tissue, even though it is not as strong as in a crystal. These effects will cause direct or indirect shifts and broadenings in the scattered light.

The absorption of light is wavelength dependent and varies greatly. For example, the main absorbers are water and chlorophyll in animal cells and vegetable cells, respectively. Of course, the absorption spectrum mainly depends on the absorbing medium.

Water exhibits good absorption in the IR region; despite this, it has a window at around 0.6 μm. Therefore, the He–Ne laser is convenient as it avoids having the signal absorbed by water. Chlorophyll absorbs in the visible range. One peak of chlorophyll absorption coincides with laser wavelength (633 nm). This helps in measuring the amount of chlorophyll in an object.

When measuring speckles, a strong absorption does not cause major problems.

However, when measuring an object where the quantity of chlorophyll changes with time, changes in speckle size may occur as less absorption involves deeper penetration of light.[37] It is possible to get rid of this problem using a diaphragm because this method keeps the speckle size constant.

Most of the absorbed energy will turn into heat, but the change of temperature will probably not be important even though there is strong absorption. The energy increase in the object depends on the experimental setup configuration. For typical conditions employed in experiments, such as a low-power laser (for example, 10 mW He–Ne) expanded beam even for several seconds of exposition, the increase in temperature is negligible.[39]

2.4 EXAMPLES OF BIOSPECKLE PHENOMENON IN NONBIOLOGICAL SAMPLES

The boiling speckle phenomenon is also present in some nonbiological processes. There are several examples where this effect is observed, including the drying of paint, corrosion, heat exchange, evolution of foams, hydroadsorption in gels. In these cases, biospeckle could have different origins and features not yet present in biological samples: for example, surface water and solvent evaporation in drying of paints, rupture and redistribution of bubbles in the evolution of the foams, surface deterioration in corrosion, salt crystallization in stone materials, etc.

In chapter 9 some cases will be commented in more detail.

2.5 BIOSPECKLES

When speckles are created from light scattered by moving particles, they are modulated by the state of motion of the scatterers. Biospeckles are very complex phenomena. When light penetrates an object, it is multiple scattered in all possible directions before leaving the object. Consequently, it is not possible to recognize and study each particle that scatters the light. In each point in the speckle image, the light from many scatterers is superposed and added on a complex basis. The resulting amplitude is

$$A(P,t) = \sum_{j=1}^{N} \left| A_j(P,t) \right| \exp[i\phi_j(P,t)] \qquad (2.8)$$

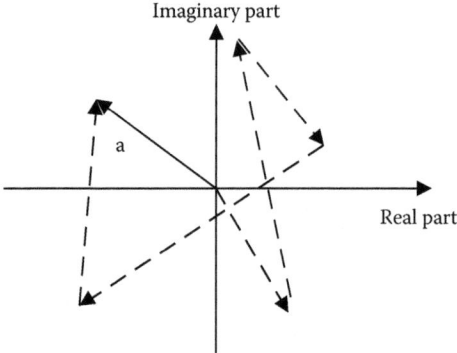

FIGURE 2.5 The sum of scatterers to create a speckle pattern.

where P is the observation point, N is the number of scatterers that contribute to A in P, and ϕ is the phase. Compare this with the complex addition in Figure 2.5; this is identical to the random walk problem in two dimensions.[3] The frequency of the speckle fluctuations are hence directly related to the rate of change in the random walk and, therefore, the velocity of each scatterer. The effect can be compared with the Doppler method where a very narrow laser beam illuminates a vessel, and the velocity (v) of the fluid in the vessel is given from the frequency of the fluctuating scattered light:

$$f_D = \frac{1}{2\pi}(\overline{K}_s - \overline{K}_i) \cdot v \tag{2.9}$$

where K_s and K_i are the wave vectors for the incident and the scattered light, respectively. In our case, every scatterer has this Doppler phenomenon, but the directions of the movements are random. Because of the complexity of the scattered light, accurate mathematical methods have not yet been established to describe this situation. However, the different wave vectors, velocity vectors, and the multiple scattering all together may statistically be interpreted as a Doppler broadening of the scattered light.[42] A more in-depth discussion will be presented in Section 2.7.

2.5.1   STATISTICS OF BIOSPECKLES

The speckle is a stochastic effect and can therefore be described only statistically.

There are many statistical approaches, and several of them are evaluations from the concepts of first- and second-order statistics. The first order is the statistical description of the intensity and contrast of speckles in a pattern in one single image. This approach shows that the standard deviation σ of the spatial intensity is the mean intensity $\langle I \rangle$ of the speckle pattern. The contrast may be expressed as $C = \frac{\sigma}{\langle I \rangle}$, and when this ratio is 1, the pattern has maximum contrast and is therefore fully developed. This, of course, depends on the time the image is exposed. With a very long exposure, the image will become unclear, resulting in very low contrast. However, with the knowledge of the time, contrast, and ratio of the moving scatterers to stationary ones, it is possible to determine the mean velocity of the scatterers. This technique can be used, for example, to visualize blood flow in the retina[5] and to study vibrations.[6]

The second-order statistics is a description of the size and intensity distributions of the speckles in an image. The method involves taking the autocorrelation of the image to determine the sizes. This technique is also used to determine the changes of intensity in one single pixel by taking the autocorrelation of its time history. This latter technique will be used in this book, and it is therefore described more thoroughly later.

In Chapter 4, the statistical properties of biospeckles will be discussed.

2.5.2 PROPERTIES OF BIOSPECKLES

The interesting dimensions of speckles are their size and intensity. The size of the stationary speckles is, as mentioned in Chapter 1, equal to the Airy disc resulting from the size of the illuminated area, but what happens to the size as the light penetrates the object and is multiple scattered? Studies have been conducted to investigate the properties of time-varying speckles. It has been shown that the speckles resulting from scattering inside an object have a smaller average size than the ones produced from scattering on the surface.[37] This can easily be explained with the expansion of the laser beam as it penetrates the object. As the light is scattered back, it leaves the object through a larger area than where it entered. In other words, the Airy law is still valid, but with the actual size of the total illuminated area rather than the width of the laser beam.

The speckle pattern is actually the superposition of two different patterns. Large speckles, resulting from scattering on the surface, and with a large angular dependence, are modulated by small speckles from the light from the interior, with very weak angular dependence.[43]

In experiments[37] made with apples, the ratio of the sizes of the speckles was found to be 10:1. The same researchers by using fixed apertures showed that the speckle size increases with the decreasing diameter of the aperture and also that the rate of temporal changes of the speckles decreases with decreasing aperture size. The time-varying effect of the speckles is also stronger far away from the direction of specular reflection.[37]

Another property that was found was the size of the region reached by light in the object when illuminating with a very thin laser beam. It has the following dependence on the depth:

$$D = D_0 \left[1 - \exp\left(-\frac{T}{T_0} \right) \right] \tag{2.10}$$

where D is the diameter of the region, and T is the depth. In the experiments with apples,[37] $D_0 = 20$ mm and $T_0 = 1.6$ mm were empirically found. A depth of about 7 mm was found to be the maximum depth from where the scattered light contributes to the speckle pattern. It was also discovered that the temporal speckle variations come mainly from the scatterers within a few millimeter depths.

2.6 BIOSPECKLE DESCRIPTION BY HOLODIAGRAM SCHEME AND SENSITIVITY VECTOR

A deeper interpretation of biospeckle phenomena requires the Doppler effect to describe the scattered waves in the samples. In the next sections we describe the Doppler effect, employing the sensitivity vector using the holodiagram.

2.6.1 SENSITIVITY VECTOR

The sensitivity of holographic and speckle measurements is governed by the orientation relation between three vectors, called the illumination vector, the displacement vector, and the viewing vector. Similar analysis could be made on biospeckle measurements. Figure 2.6 shows these vectors in a general measurement situation.

The object is illuminated by a point source located at S. Light is scattered by an object point P to an observer at point O. In our case, the object is a single particle (or scatterer) of the sample. When the object is moved from the point P to point P' by the displacement vector $\mathbf{d}$, there is a change in the optical path from the point source S to point P' and from there to the viewing point O. So, there is a change in phase associated with this change in optical path. This phase difference is due to two reasons: the scatterer changes its distance from the source, and it also changes its distance to the detector.

The phase shift $\Delta\phi$ of light scattered by the object when the point P moves to P' is given by

$$\Delta\phi = \frac{2\pi}{\lambda}(\mathbf{n}_1 + \mathbf{n}_2)\cdot\mathbf{d} = \frac{2\pi}{\lambda}\mathbf{S}\cdot\mathbf{d} \qquad (2.11)$$

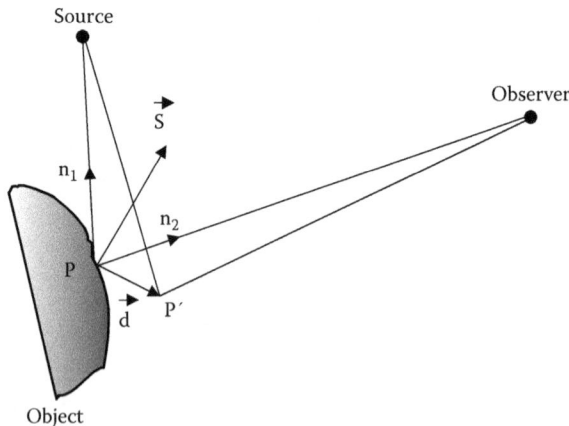

FIGURE 2.6 Sensitivity vector schema: n_1 and n_2 are the unitary vectors in the illumination and viewing directions, respectively, $\mathbf{S} = (\mathbf{n}_1 + \mathbf{n}_2)$ is the sensitivity vector, $\mathbf{d}$ displacement vector.

where λ is the illuminating wavelength, and $\mathbf{n}_1$ and $\mathbf{n}_2$ are the unitary vectors in the illuminating and viewing directions, respectively. $\mathbf{S} = (\mathbf{n}_1 + \mathbf{n}_2)$ is named the *sensitivity vector.*

The result shows that, if the object and illumination beam directions are known and if the phase change is measured, a component of the displacement vector can be determined. As expected, if the displacement has three unknown components, then three observations with different viewing directions will be required, and there will be three equations to solve simultaneously.

The sensitivity vector $\mathbf{S}$ has the following physical meaning. Let θ be the angle between the illumination and the viewing beam directions. Then, the magnitude of $\mathbf{S}$ is $2\cos(\theta/2)$ and it is directed along the bisector of the angle between the illuminating and viewing directions. This fact can be helpful in establishing the best directions for measuring a particular displacement component.

2.6.2 HOLODIAGRAMS

The sensitivity vector concept can be better understood by means of a useful scheme called the Holodiagram.[28] It consists in drawing the loci of points showing the constant sum of the distances between a generic point P and two other points, figuring a light source A and an observation point B. In free space, it consists of a set of revolution ellipsoids. Its plane cuts are ellipses. Consecutive ellipses differ, by definition, in their parameter by half a wavelength.

If a particle crosses from one ellipse to the next, its path length increases or decreases by half a wavelength. The detector in B is assumed to be phase sensitive, a feature that in holography is achieved by adding a reference beam, and in the dynamic speckle by the presence of other coherent waves coming from other scattering centers.

Figure 2.7 shows the holodiagram and the sensitivity vector $\mathbf{S}$ for point $\mathbf{P}$.

If P displaces to P′, light coming from A and going to B changes its phase according to Equation 2.11.

That is, the phase change is proportional to the component of $\mathbf{d}$ in the direction of $\mathbf{S}$.

$$\Delta\phi = \frac{2\pi}{\lambda}\mathbf{S}\cdot\mathbf{d} = \frac{\pi\mathbf{S}\cdot\mathbf{d}}{(\lambda/2)} \qquad (2.12)$$

That is, when $\mathbf{S}\cdot\mathbf{d}$ is measured in half wavelengths, the result is the phase measured in π units. Now, $\mathbf{S}\cdot\mathbf{d}/(\lambda/2)$ is the number of ellipses crossed by the point P. So, this number represents the phase change in π units. The phase change can be estimated by counting the number of ellipses that the point P crosses when moving. If it does not cross any, that is, if it moves within a certain ellipse, there will be no phase change. If the movement is in a direction where the ellipses are densely packed, it will produce a high phase change. Thus, it can be seen that the diagram is useful to estimate the sensitivity to movements of the scattering centers.

The points along one of the ellipses have constant optical path length. If a scattering particle moves along one of the ellipses, it will not produce any measurable effect in the light detected at point B. Conversely, regions of the holodiagram where

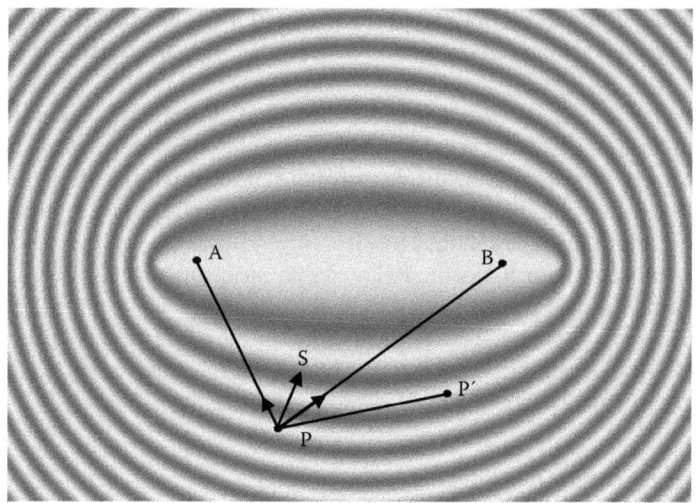

FIGURE 2.7 The sensitivity vector analysis using holodiagram schema.

its lines are densely packed will be highly sensitive to the scattering center's movement in the direction perpendicular to the ellipses.

Notice that, in the region between A and B (corresponding to forward scattering), there is a (comparatively) large region where sensitivity to displacements in any direction is low. On the other hand, regions at the left of point A and the right of point B show high sensitivity to movement in the direction defined by the AB segment (corresponding

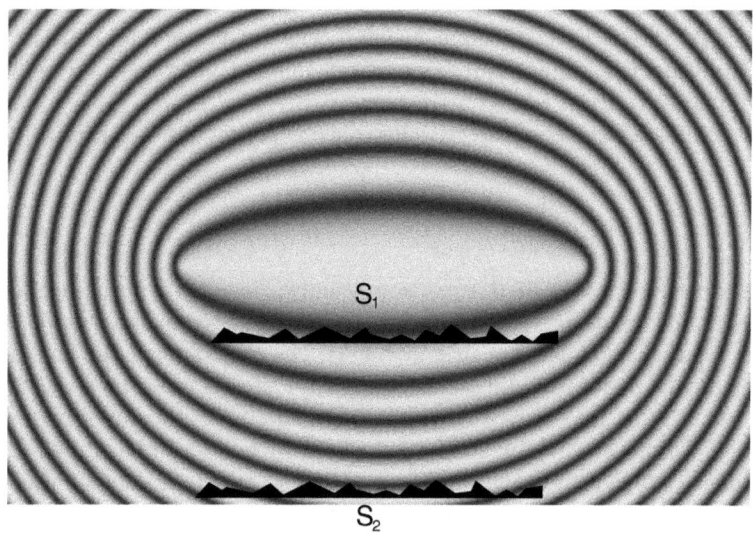

FIGURE 2.8 Two locations of the scatterers. The surfaces are illuminated from A and observed from B.

to backscattering). By looking at the holodiagram, a fast estimation of the comparative sensitivity of the different geometries of illumination–observation is obtained.

Now, let us consider the analysis of the speckle phenomenon using this approach.

An object optically rough, or a diffuser S, is illuminated by a light source in A and is observed by a detector in B. Figure 2.8 shows the object located in position S_2 whose profile cuts several ellipses, which means that phase excursions are greater than π, giving rise in B to a fully developed speckle pattern. When the same object is in position S_1, the phase excursions are smaller (in some regions they do not cut consecutive ellipses); consequently, the speckle pattern in B may not be fully developed.

2.7 DOPPLER EFFECT

The frequency of photons scattered from moving particles will be modified depending on the speed of the moving parts. An illustration of the effect is the sound of cars on the highway changing in perceived frequency between coming toward you and going away from you. This is called the Doppler effect, which is closely related to the process in the time-varying speckles.

The Doppler effect can be analyzed using the concept of holodiagram developed in Section 2.6. A similar situation as in Figure 2.8 may happen with the movement of the scattering centers. If the object is not rigid but is constituted by a set of discrete points, such as gas particles, each one randomly moving, all the described situations may be present at the same time.

Figure 2.9 illustrates the holodiagram of this situation in a general case where several particles of the scattered object are considered. As said earlier, A is the point source, and B is the observation point. As a particle P with a velocity v_p passes through the ellipses of the holodiagram, there will be a set of light pulses on the

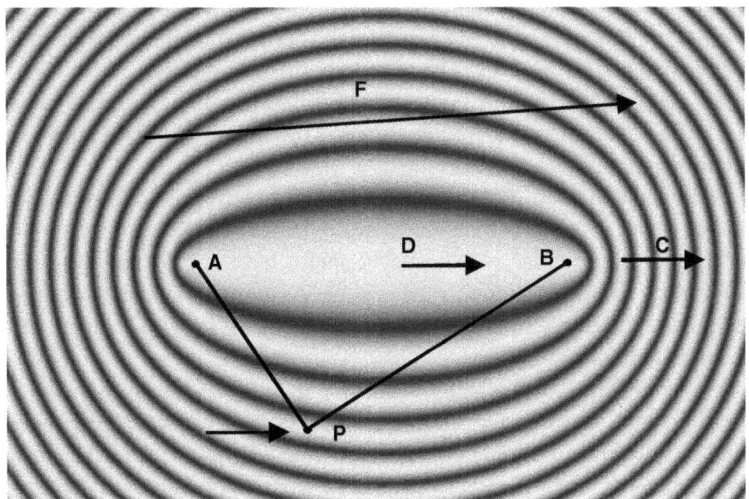

FIGURE 2.9 Doppler effect analysis using a holodiagram. Several movements of particles in different directions are illuminated from A and observed from B.

detector, one pulse for every time an ellipse is crossed. The direction of zero sensitivity is parallel to the ellipses, whereas the maximum sensitivity is perpendicular to the ellipses. The frequency of the pulses is therefore a measure of the particles' velocity. The velocity v is calculated by

$$v = f k \,(\lambda/2) \qquad\qquad (2.13)$$

where f is the frequency of the signal in the detector, $k = \frac{1}{\cos(\theta/2)}$ is called the k value of the holodiagram that is proportional to the separation of the ellipses, θ is the angle between the illumination and viewing directions, and λ is the wavelength of the illumination light.

Where the k value is 1 in the direction along the x-axis outside the AB line, there is maximum sensitivity; whereas where the k value is infinite, the sensitivity is zero on the AB line.

Let us consider particles with C, D, E, and F velocities, and f as the frequency measured in B. Particles of C velocity produce high f, whereas D and E velocity particles produce zero f. The particle of F velocity produces a signal with f varying from high to zero, f and returning again to high f.

Note that high or low sensitivity may be good or bad depending on the circumstances. Low sensitivity may be required, for example, for a very active phenomenon to be resolved by a comparatively slow detector.

This analysis demonstrates that the holodiagram is a useful tool to interpret the signal on the detector produced by a boiling speckle pattern. The observed frequency on the detector has the expression of the Doppler effect. So, the origin of a boiling speckle pattern can be considered as a Doppler effect due to the random movement with different velocities of the particles in the sample.

The range of speeds in the particles that contribute to the formation of the biospeckle is very wide. For example, the average velocity of the particles in the apple[37] is 0.4 μm/s; in the orange it is 1.8 μm/s; in human sperm[44] it fluctuates from 66 μm/s (T = 24°C) to 165 μm/s (T = 37°C). The microorganism *Spirillum serpens* turn their flagella at a speed of 2400 rpm; its body speed is of 800 rpm, moving at a velocity[45] of 50 μm/s; *Vibrio cholerae* is moving at a speed[45] of 200 μm/s; and the mean velocity of scatterers in paint[10] is around 10 nm/s.

It can be thought that, in the speckle pattern, fluctuation is mainly caused by interference of many rays rather than shifts in single photons. According to the central limit theorem, independent stochastic variables have approximately a normal distribution as long as the number of trials is large. Assuming that the random movements of the many scatterers are independent, they obey the normal distribution, and hence, the Gaussian one.[46]

In this case, unlike the Doppler method, the directions of the movements of the scatterers are of no consequence to the speckle pattern. Both effects cause the same frequencies and broadenings, and they both appear in time-varying speckles. The intensity fluctuation is, however, more dominated by the random phase variation due to interference rather than by the Doppler shift of photons, and therefore it can be preferably called time-varying speckle phenomenon. However, in the Section 2.9, we follow the analysis of Briers,[29] where he demonstrates the equivalence of both points of view.

2.8 BEATING PHENOMENON

Assuming that the biospeckle is basically a Doppler effect phenomenon, there will be many scatterers moving at different velocities, and there will also be many associated frequencies. These frequencies correspond to the laser frequency with very small shifts, and so they cannot be resolved by the detector. Therefore, we must also consider the phenomenon of *beat frequency*.

In the simplest one-dimensional case, we consider the superposition of two waves, a_1 and a_2, with the same amplitude A and slightly different angular frequencies, β_1 and β_2, respectively.

$$a_1 = A \exp (i\beta_1 t) \tag{2.14}$$

$$a_2 = A \exp (i\beta_2 t) \tag{2.15}$$

Considering the superposition principle, the resulting wave is

$$a = a_1 + a_2 = 2A \exp [i((\beta_1 - \beta_2)/2) t] \exp [i((\beta_1 + \beta_2)/2) t] \tag{2.16}$$

So, the resulting wave has an average angular frequency between the two waves, and the amplitude slowly varies in time at an angular frequency given by $(\beta_1 - \beta_2)/2$, as shown in Figure 2.10. In principle, this frequency can be resolved by ordinary detectors, provided $(\beta_1 - \beta_2)$ is small enough.

In the case of light, the corresponding intensity $I = a\,a^*$ varies slowly and constitutes the most simple case of beating that is the base of the time-varying speckle phenomenon. Here, many overlapping waves originate in the movement of the scatterers. Therefore, Equation 2.16 must be generalized.

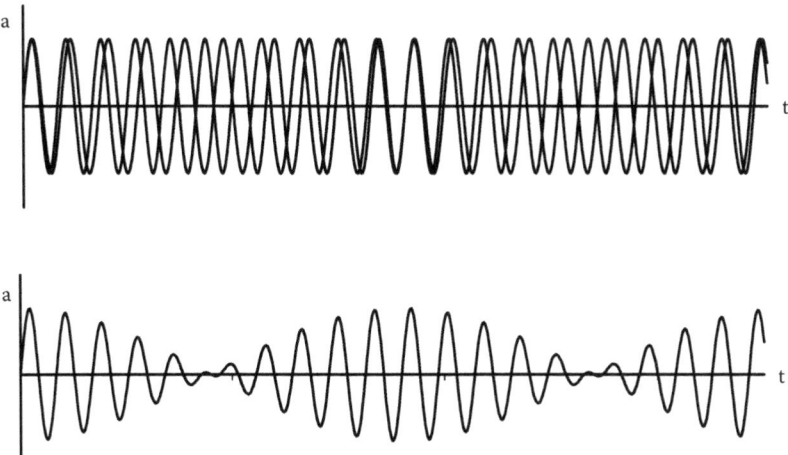

FIGURE 2.10 Beating phenomenon: (a) two waves a_1 and a_2 with the same amplitude A and slightly different frequencies β_1 and β_2, respectively; (b) resulting wave $a_1 + a_2$.

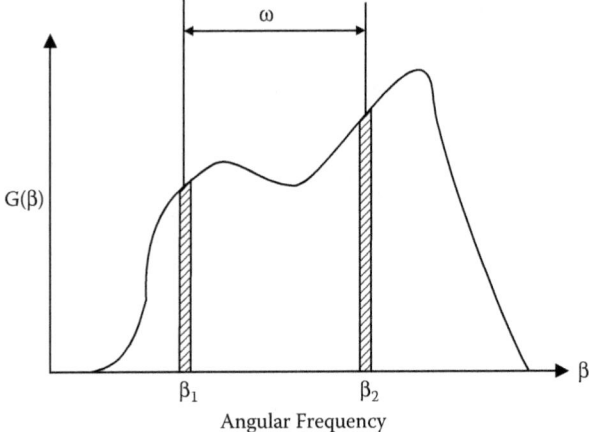

FIGURE 2.11 $G(\beta)$ is the power per unit bandwidth of light as a function of angular frequency β.

According to Forrester,[26] to derive the intensity registered by the detector device, we must consider all the contributions coming from all the moving scatterers that produce different frequencies.

In Figure 2.11, $G(\beta)$ represents the power per unit bandwidth of light as a function of angular frequency β of an active source. Two frequencies, for example, β_1 and β_2, beat between them to produce a frequency $\omega = \beta_1 - \beta_2$ in the detector.

We divide the frequency range into a large number of narrow bands of $\Delta\beta$ width. The intensity due to the m-th interval is given by

$$(\Delta I)_m = G(\beta_m)\Delta\beta \tag{2.17}$$

The corresponding electric field amplitude is

$$E_m = (G(\beta_m))^{1/2} \exp(j(\beta_m t + \phi_m)) \tag{2.18}$$

The intensity is obtained as

$$I = E E^* \tag{2.19}$$

The phase ϕ_m is randomly taken with respect to all other phases. The total field amplitude is obtained by adding all m values. Then the intensity is

$$I = \Delta\beta \left[\sum_{m=1}^{\infty} (G(\beta_m))^{1/2} \exp(j(\beta_m t + \phi_m)) \right] \left[\sum_{m=1}^{\infty} (G(\beta_m))^{1/2} \exp(-j(\beta_m t + \phi_m)) \right] \tag{2.20}$$

And, after changing one of the indices, this can be written as

$$I = \Delta\beta \sum_m \sum_n [G(\beta_m)G(\beta_n)]^{1/2} \times \exp j[(\beta_m - \beta_n)t + (\phi_m - \phi_n)] \qquad (2.21)$$

The DC terms are obtained when $m = n$, and are

$$I_0 = \Delta\beta \sum_{m=1}^{\infty} G(\beta_m) \xrightarrow[\Delta\beta \to 0]{} \int_0^{\infty} G(\beta)d\beta \qquad (2.22)$$

In Equation 2.20 we can group all terms with a single frequency by changes in one of the indices, say, m to $b = m - n$.

$$I = \Delta\beta \sum_{n=1}^{\infty} \sum_{b=1-n}^{\infty} [G(\beta_n)G(\beta_n + \omega_b)]^{1/2} \exp j(\omega_b t + (\phi_{b+n} - \phi_n)) \qquad (2.23)$$

where $\omega_b = b\,\Delta\omega$. This sum extends over both positive and negative values of b corresponding to $m<n$ and $m>n$. Changing the order of summation,

$$I = 2\Delta\beta \sum_{b=1}^{\infty} \sum_{n=1}^{\infty} [G(\beta_n)G(\beta_n + \omega_b)]^{1/2} \times \cos(\omega_b t + \phi_{b+n} - \phi_n) \qquad (2.24)$$

where $b = 0$ terms (the DC current) have been omitted.

After considering that the detector can only follow signals in a small frequency interval $(\omega, \omega + \Delta\omega)$, Forrester finds for the mean square value of the current

$$\langle I^2(\omega) \rangle = 2\Delta\omega \sum_{n=1}^{\infty} G(\beta_n)G(\beta_n + \omega)(\Delta\beta) \qquad (2.25)$$

that, when $\Delta\beta \to 0$, results in

$$\langle I^2(\omega) \rangle = 2\Delta\omega \int_0^{\infty} G(\beta)G(\beta + \omega)d\beta \qquad (2.26)$$

Then, the root-mean-square (RMS) current measured by the detector is proportional to the autocorrelation of $G(\beta)$.

2.9   COMPARISON BETWEEN BIOSPECKLE AND LASER DOPPLER

Coherent laser light scattered from moving objects or particles produces intensity fluctuations. Two approaches have been used to analyze these fluctuations: laser Doppler and time-varying speckle. These two techniques have developed separately, and it is not evident that they are effectively identical.

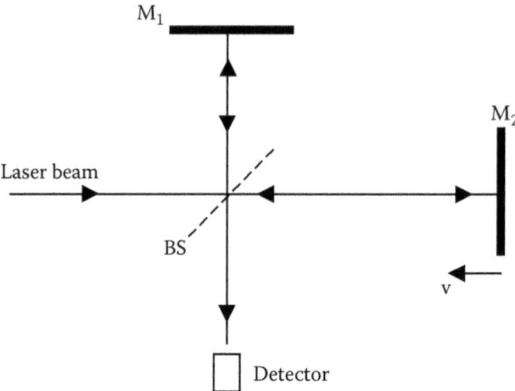

FIGURE 2.12 Michelson interferometry. M_1 is fixed, and M_2 moves with velocity v.

Briers[29] demonstrated the equivalence between the two techniques. We follow his approach in the next sections.

2.9.1 EQUIVALENCE OF DOPPLER AND INTERFEROMETRY

Consider the Michelson interferometer, where one of the mirrors, M_1, is stationary, and mirror M_2 can move with velocity v along the direction of the light beam, as shown in Figure 2.12. When M_2 moves, a detector D located at the output of the interferometer records a sinusoidal fluctuating intensity signal. The velocity of M_2 is very low in comparison with the light velocity c.

2.9.1.1 Doppler Interpretation

The velocity of the wave of the incident beam is c, and its frequency is f. If M_2 is stationary, f wave crests occupy a distance **c** in one second . On the contrary, if M_2 moves toward the light source at a velocity v ($v \ll c$), the wave crest occupies, after reflection, a distance of $c - 2v$ per second. Then, the frequency of the reflected wave on the detector D is

$$f' = \frac{c}{c - 2v} f \tag{2.27}$$

At D the reflected wave of frequency f' is mixed with the frequency f of the local oscillator (reference wave), and the beat frequency detected is

$$\delta f = f' - f = f \left[\frac{c}{c - 2v} - 1 \right] = f \frac{2v}{c - 2v} \tag{2.28}$$

As $v \ll c$

$$\delta f = f \frac{2v}{c} \tag{2.29}$$

Equation 2.29 gives the Doppler frequency recorded by the detector D. As c and f are known, the velocity v of M_2 can be calculated.

2.9.1.2 Interferometric Explanation

The reflected beam from M_2 interferes with the beam from M_1. The intensity in the detector plane D is uniform if both mirrors are flat and located perpendicularly to the light beams. Conversely, if the mirrors are slightly misaligned, a fringe pattern can be observed in the plane detector, depending on the phase difference.

If M_2 moves a distance equal to $\lambda/2$, where λ is the light wavelength, the optical path difference between the two beam changes in λ. Then, detector D records a complete cycle of interference.

If δf cycles are recorded in 1 s, it implies that M_2 has moved a distance v in that second, and

$$v = \delta f \frac{\lambda}{2} \tag{2.30}$$

In this case, v is the velocity of M_2, and δf is the frequency of the intensity oscillations recorded by D.

Using the relationship $c = f\lambda$ and rearranging Equation 2.30 gives

$$\delta f = f \frac{2v}{c} \tag{2.31}$$

which is identical to Equation 2.29.

It is by no means intuitively obvious that the two models are identical. The Doppler explanation involves the superposition of two waves of slightly different frequencies and the detection of the resulting beat frequency. The interference explanation involves the superposition of two waves of the same frequency and the detection of correlations when one wave slides past the other. Nevertheless, the previous analysis demonstrates that the detected frequency is the same in both cases. The two approaches are in fact two ways of looking at the same phenomenon.

2.9.2 Equivalence of Doppler and Time-Varying Speckle

If mirror M_2 is replaced by a rough diffusing surface, the light scattered from it produces a far-field speckle pattern in the detector plane. This speckle pattern is still coherent with the reference beam and interferes with it to produce a new speckle pattern. A Doppler signal is still produced by detector D, and the velocity of the moving diffuser is still given by Equation 2.29. Similarly, the interference interpretation is unaffected. The intensity at any point in the speckle pattern fluctuates if M_2 moves and again passes through one cycle for each half-wavelength movement of M_2. Thus, Equation 2.31 is still valid.

If M_2 consists of a collection of individual scattering particles rather than a solid object, the same reasoning is pertinent, except for the existence of a range of velocities associated with the scatterers. It results in a spectrum of frequencies instead of a single frequency that is detected. Both approaches, the Doppler and the time-varying speckle, are suitable for explaining the phenomenon and give the same quantitative answers.

2.9.3 PHOTON CORRELATION SPECTROSCOPY, DOPPLER, AND TIME-VARYING SPECKLE

Now consider the situation without the reference beam. A set of scatterers that are moving independently with varying velocities is illuminated with coherent light, and the intensity is monitored at a point in the far field of the scattered light. Intensity fluctuations are still observed. In interference terms, the intensity at D is determined by the coherent addition of the waves scattered from all the illuminated particles. There is no reference beam to interfere with this time, but as the path lengths from the different particles vary, the resultant complex amplitude at D, and hence the intensity, also vary in time. The result is a time-varying speckle pattern.

In Doppler terms, the situation is now a homodyne one rather than the heterodyne case when the reference beam (local oscillator) is present. Different frequencies of the light scattered from particles with different velocities beat with each other to give a Doppler spectrum. However, note that the loss of the reference beam means that information about the global velocity of the set of scatterers is lost: the Doppler spectrum merely gives the spread of velocities about an unknown mean velocity. For this reason, the homodyne technique is usually used in situations in which there is no overall global velocity but in which the parameter of interest is the actual spread of velocities. Examples include the studies of diffusion rates, Brownian motion, and motility of biological organisms.

Traditionally, these situations have been investigated by measurement of the autocorrelation function of the intensity fluctuations rather than the frequency spectrum and the technique has been known as intensity fluctuation spectroscopy, photon correlation spectroscopy, or light-beating spectroscopy.[27]

The autocorrelation function and the frequency spectrum form a Fourier transform pair, so this distinction between photon correlation spectroscopy and Doppler is merely for convenience rather than anything fundamental. The former is used for the homodyne case (when the correlation function is measured), whereas Doppler is reserved for the heterodyne technique (when it is usual to measure the frequency spectrum).

The conclusion from this analysis is that there is basically no difference among the phenomena of time-varying speckle, laser Doppler, and photon correlation spectroscopy. There is a difference of approach and sometimes a different parameter may be measured, but fundamentally they represent the same technique.

2.10 REFRACTIVE INDEX AND CONFIGURATIONAL EFFECTS

2.10.1 REFRACTIVE INDEX

Refractive index time variations can be due to diffusion phenomena producing concentration changes or movements of liquids. Turbulence could be included in this type of contribution. The twinkling of the stars, a stellar speckle time variation, is an example where the time-varying speckle is associated with the time dependence of the local refractive index and not with the velocities of individual scatterers. In the general case, when the scatterer moves and the refractive index changes with time, which is the rule more than the exception in biospeckle experiments, both effects must be taken into account. Moreover, in a biological phenomenon, space and time refractive index variations can be expected to be partially correlated.

Laser Doppler line-of-sight velocity measurement techniques and speckle interferometry have been shown to be two different viewpoint descriptions of the same experiments.[29] For the description of dynamic speckle patterns in terms of Doppler shifts to be fully equivalent to that in terms of time-dependent interference, the effect of refractive index time variation and configurational effects should be taken into account.

When this is so, the space- and time-dependent variables appear multiplied, and several space concepts, such as that of (surface) roughness, can be extended to the time variable to define a time diffuser. Then, the cross-fertilization between the concepts in the Doppler field and those of speckle optics can be made more complete. Some roughness measurement methods could benefit from being translated into the time domain to include these phenomena. Digital speckle pattern interferometry (DSPI) experiments are in one extreme of the spectrum of speckle experiments that can be considered as a conceptual whole. Configurational effects and the possibility of their time change should also be included.

In the previous section, Briers[29] showed that laser Doppler and speckle interferometry techniques were equivalent approaches when used to measure the line-of-sight velocities of scatterers. He used a Michelson interferometer as an example to illustrate his description. He appropriately described the dynamics of the moving scatterers usually found in the Doppler effect.

However, one aspect is not explicitly included in Brier's description, and it may be of importance when those approaches are used in the time-varying speckle patterns produced by biological samples, among others. When the scatterers are still but the medium changes its properties, the speckle pattern changes with time in certain ways that cannot be discerned from the Doppler approach, but in fact can be considered a manifestation of its effect.

Let us use the same paradigm as Briers (but for a different reason); it is, the Michelson interferometer. Figure 2.13 shows a classical interferometer with equal arms, except for the introduction of a cuvette containing a liquid of refractive index n.

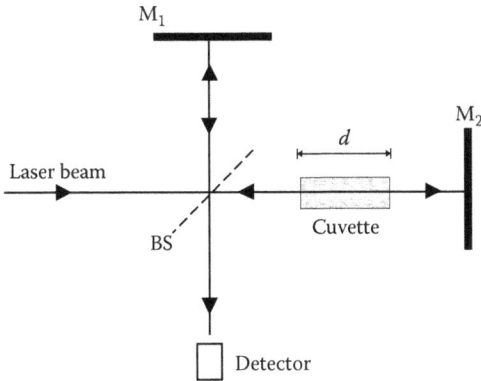

FIGURE 2.13 Michelson interferometer with a cuveter of index n in one of the arms.

If n is permitted to change with time in a certain assumed, continuous way $n(t)$ by mixing the liquid with another having a different refractive index, then the resulting interference intensity will also show time variations.

Let us assume that the refractive index variations are very slow compared with the integration time of the detector. The detected intensity at the output will be

$$I = 4I_0 \cos^2 \frac{\delta(t)}{2} \qquad (2.32)$$

where I_0 is the intensity that will be measured if only one arm of the interferometer is present, and $\delta(t)$ is the phase difference caused by the difference in optical paths between both arms.

Taking into account the cuvette thickness d and refractive index of the liquid $n(t)$, $\delta(t)$ can be expressed as

$$\delta(t) = \frac{4\pi}{\lambda_0}[n(t) - 1]d \qquad (2.33)$$

where λ_0 is the vacuum wavelength of the light.

It can be seen that the detector will show intensity variations bright–dark–bright as in the case of the moving mirror in the Briers[29] approach, but here due to refractive index variations. The "instantaneous" frequency[47] of the detected signal will be

$$\Delta f = \left(\frac{1}{2\pi}\right)\frac{d}{dt}\delta(t) = 2\frac{d}{\lambda_0}\frac{d}{dt}n(t) \qquad (2.34)$$

That is, by only using the detector, the change in the refractive index of the liquid in the cuvette cannot be distinguished from the movement of a mirror with a line-of-sight velocity v given by

$$v = d\frac{d}{dt}n(t) \qquad (2.35)$$

Thus, the observed intensity in the detector is the same in both cases: the movement of the particles or a small change in the refraction index.

If we consider now a slightly more general experiment where both the refractive index n and the cuvette thickness d are changing in time, and we call

$$v = \frac{d}{dt}d(t) \qquad (2.36)$$

then

$$\Delta f = \left(\frac{1}{2\pi}\right)\frac{d\delta(t)}{dt} = \frac{1}{\lambda_0}\left[n(t)\cdot v + d\frac{dn(t)}{dt}\right] \qquad (2.37)$$

The observable quantity is dependent on the time derivative of an optical path where $n(t)$ and $d(t)$ appear multiplied, and a single measurement cannot distinguish between the contribution of the velocities of scatterers and the time change of spatial properties of the medium $n(t)$ and $d(t)$.

We can generalize the experiment a little more by supposing that the whole arm of the interferometer is filled with the changing liquid and that the thickness variation of the cuvette is due to the movement of the mirror, in analogy with Briers.[29] Better still, we can even think that the mirror has been changed by a diffuser. As the product of the refractive index multiplied by the distance is involved in the phase calculation, it might be better to think in terms of a time-varying optical roughness. The refractive index and surface profile appear as a nonseparable product, and so are the frequency shifts due to time changes in both magnitudes.

2.10.2 Configurational Effects

As it is well known in elementary descriptions of the Doppler effect, the movement of the source and the change in the observation direction modify the value of the detected frequency, as the sensitivity vector also changes its direction. In the general case of the boiling speckle phenomenon, all the previously described effects are present at the same time.

If $\mathbf{d}$ is now the displacement of a scatterer in a certain direction, then the optical path between the laser source and an observation point changes because the scatterer changes its distance from both the source and the detector. Then δ is given by[48]

$$\delta(P) = \frac{2\pi}{\lambda}\left[\mathbf{n}_1(P) + \mathbf{n}_2(P)\right] \cdot \mathbf{d}(P) = \left(\frac{2\pi}{\lambda}\right)\mathbf{S}(P) \cdot \mathbf{d}(P) \qquad (2.38)$$

where here again $\mathbf{n}_1$ is a unit vector from P in the illumination direction, $\mathbf{n}_2$ is a unit vector from P in the observation direction, and $\mathbf{S}$ is the sensitivity vector. That is, the velocity component detected is in the direction of the sensitivity vector, as in holographic interferometry. This direction and other useful ones, such as the direction that does not produce observable Doppler shift, can be graphically observed in the holodiagram.

Handling the sensitivity vector in order to make an experiment conform to the present requirements is a well known and exploited maneuver in holographic interferometry and DSPI techniques. To fully determine the velocity of a point source, three sensitivity vectors are required as a minimum, unless some other information, such as the direction of movement, is available. Unfortunately, a holodiagram that includes variable refractive index effects has not been developed.

The phase derivative with respect to time gives rise to the detected frequency change. So, if the sensitivity vector is time dependent because the source or detector moves, or the geometric distribution changes, or both, these effects will be combined with the scatterer movement in a compound frequency shift. Certainly, in most experimental situations, the source and the detector are kept as still as possible, but in biospeckle, a scatterer can receive light contributions from another moving one, the latter acting as a source and further as a detector (multiple scattering).

The developed analogy, which suggests that all these phenomena can be thought of as different variations of the Doppler effect, should include the possibility of a change in the position of the source and observation points, as in conventional Doppler effect as well as the time variation of the refractive index $n(t)$ and how this variation interacts with the other terms. As the measurements are made on the intensity, there is no way to distinguish if a variation is due to a movement in the mirror, a change in the refractive index, a change in the configuration, or because a bubble has exploded. These effects are added and could eventually partially compensate each other.

2.11   OTHER PHENOMENA IN BIOSPECKLE ORIGIN

The frequency of photons scattered from moving particles will be modified depending on the speed of the moving parts, but biospeckle fluctuation also depends on the phase changes of many different waves of light that interfere to produce the biospeckle pattern. As said earlier, the biospeckle can be described in terms of the Doppler effect and frequency variations, but others effects as turbulence and diffusion can also be present and must be considered.

The Doppler effect obeys Gaussian statistics. According to the central limit theorem, independent stochastic variables have an approximately normal distribution as long as the number of trials is large. Assuming that the random movements of many scatterers are independent, they obey the normal distribution and hence the Gaussian one.[46] Turbulence is a problem that occurs, for instance, in astronomy when studying the atmosphere optically. Warm and cold air have different refractive indexes, and the differences in pressure changes the index as well. A similar phenomenon exists in biological tissue. Several models have been created to statistically solve these kinds of problems. Lee and Harp[49] presented a model in which the medium is imagined to be consisting of layers of temporally and randomly different refractive indexes.

Every layer will shift the light through a certain time-varying phase. Because the different planes act independently, and the phase fluctuations within each plane obey the same physical laws, the net optical phase in the detector satisfies the central limit theory and is therefore normally distributed. The normal distribution has a Gaussian line shape and is therefore superposed with the Doppler broadening to another Gaussian one.

The fluids in and between the cells are mixtures of different liquids. They do not propagate but have fluctuations in density and concentration. The relaxation of these fluctuations is diffusive. As in any other matter, thermodynamics always oscillates around equilibrium, and because biological matter is alive, this process is continuous.

The result of the fluctuations in the concentration and density of the sample has a Lorentzian line shape. Of course, there are many different diffusion processes in the biological tissue, resulting in a series of Lorentzians with different line widths, all centered at f_0 with an added total line width.

The Voigt profile[50] is a convolution of a Gaussian line shape and a Lorentzian line shape, which usually appears in optics and spectroscopy researches. One can think of it as if every single Doppler element (Gaussian profile) were smeared out into a Lorentzian profile. Actually, there are often several different Lorentzian and Gaussian contributions to the shape. The greatest difference between the two shapes is that the Lorentzian profile has wider wings than the Gaussian one. This results in the wings of the Voigt profile being almost fully decided by the Lorentzian one irrespective of its contribution.

Recently, D. D. Duncan and S. J. Kirkpatrick[51] used the concept of a copula to generate Gaussian and exponential forms for the decorrelation of an objective speckle sequence. In the case of both ordered and unordered motion of the scatterers, the exponential corresponds with Lorentzian temporal power spectrum. The product of the exponential and Gaussian forms corresponds with a convolution of the line profiles, producing a Voigt profile.

2.12 CONCLUSIONS

The phenomenon of speckles that changes over time is correlated to the change of physical quantities on the sample illuminated with a coherent light, a variation of the refractive index, physical position, scattering, etc. It is characteristic of biological samples and is named biospeckle. The visual appearance of the speckle diagram is similar to that of a boiling liquid.

In this chapter we discussed some concepts related to the origin and features of the biospeckle phenomenon. Light interacts with matter and can be reflected, scattered, absorbed, or reradiated when it interacts with bioobjects. So, these phenomena are the initial cause of the formation of biospeckle patterns. Also, in nonbiological cases, biospeckle patterns are present when the samples change in time due to local movements of its particles or changes in its optical properties.

These phenomena can be analyzed in terms of the Doppler effect and light-beating spectroscopy. We used the sensitivity vector and the holodiagrams scheme for this description. Also, Briers demonstrates the equivalence between time-varying speckle phenomenon, laser Doppler velocimetry, and photon correlation spectroscopy.

Other aspects, such as the effects produced by changes in the refractive index of the samples, and configurational changes in the experimental setup, the light source, or in the detectors were analyzed.

Turbulence and diffusion that introduce biospeckle fluctuations are also discussed. In these cases, Gaussian and Lorentzian spectra are obtained and analyzed by Voigt profiles.

REFERENCES

1. Dainty, J. C., *Laser Speckle and Related Phenomena*, Dainty, J. C., Ed., Springer Verlag, Berlin, 1975.
2. Sirohi, R. S., *Speckle Metrology*, CRC Press, New York, 1993.
3. Goodman, J. W., *Speckle Phenomena in Optics: Theory and Applications*, Roberts and Co., Englewood, Colorado, 2007.
4. Aizu, Y. and Asakura, T., Biospeckles, in *Trends in Optics*, A. Consortini, Ed., Academic Press, San Diego, 1996, chap. 2.

5. Briers, J. D., Speckle fluctuations and biomedical optics: Implications and applications, *Opt. Eng.*, 32, 277, 1993.
6. Eliasson, B. and Mottier, F., Determination of the granular radiance distribution of a diffuser and its use for vibration analysis, *J. Opt. Soc. Am.*, 61, 559, 1971.
7. Asakura, T. and Takai, N., Dynamic laser speckles and their application to velocity measurements of the diffuse object, *Appl. Opt.*, 25, 179, 1981.
8. Creath, K., Phase shift speckle interferometry, *Appl. Opt.*, 24, 3053, 1985.
9. Briers, J. D., The statistics of fluctuating speckle patterns produced by a mixture of moving and stationary scatterers, *Opt. Quant. Electr.*, 10, 364, 1978.
10. Rabal, H. J., Arizaga, R., Cap, N., Grumel, E., and Trivi M. Numerical model for dynamic speckle: an approach using the movement of scatterers, *J. Opt. A: Pure Appl. Opt.*, 5, S381, 2003.
11. Federico, A., Kaufmann, G. H., Galizzi, G. E., Rabal, H. J., Trivi, M., and Arizaga, R. Simulation of dynamic speckle sequences and its application to the analysis of transient processes, *Opt. Comm.*, 260, 493, 2006.
12. Okamoto, T. and Asakura, T., The statistics of dynamic speckles, *Progr. Opt. XXXIV*, E. Wolf, Ed., Elsevier, Amsterdam, 1995.
13. Amalvy, J. I., Lasquibar, C. A., Arizaga, R. Rabal, H. J., and Trivi, M. Application of dynamic speckle interferometry to the drying of coatings, *Progr. Org. Coat.*, 42, 89, 2001.
14. Muramatsu, M., Guedes, G.H., and Gaggioli, N. G., Speckle correlation technique used to study the oxidation process in real time, *Opt. Laser Tech.*, 26, 167, 1994.
15. Bandyopadhyay, R., Gittings, A. S., Suh, S. S., Dixon, P. K., and Durian, D. J. Speckle visibility spectroscopy: A tool to study time varying dynamics, *Rev. Sc. Instr.*, 76, 093110, 2005.
16. Zanetta, P. and Facchini, M., Local Correlation of laser speckle applied to the study of salt efflorescence on stones surfaces, *Opt. Commun.*, 104, 35, 1993.
17. Cabello, C. I., Bertolini, G., Amaya, S., Arizaga, R., and Trivi, M. Hydroabsorption analysis by speckle techniques, in *Proceedings of Tecnolaser 2007*, Darias, J., Ed., Electronic Edition, 2007.
18. Aizu Y. and Asakura, T., Biospeckle phenomena and their application to the evaluation of blood flow, *Opt. Laser Tech.*, 23, 205, 1991.
19. Pomarico, J. A., Di Rocco, H. O., Alvarez, L., Lanusse, C., Mottier, L., Saumell, C., Rabal, H. J., Arizaga, R., and Trivi, M. Speckle interferometry applied to pharmacodynamics studies: Evaluation of parasite motility, *Eur. Biophy. J.*, 33, 694, 2004.
20. Tearney, G. J. and Bouma, B. E., Atherosclerotic plaque characterization by spatial and temporal speckle patterns, *Opt. Lett.*, 27, 533, 2002.
21. Dunn, A. K., Devor, A., Bolay, H., Anderman, M. L., Moskowictz, M. A., Dale, A., M. Boas. D. A. Simultaneous imaging of total cerebral hemoglobin concentration, oxygenation, and blood flow during functional activation, *Opt. Lett.*, 28, 28, 2003.
22. Cheng, H., Luo, Q., Wang, Z., Gong, H., Chen, S., Liang, W., and Zeng, S. Efficient characterization of regional mesenteric blood flow by use of laser speckle imaging, *Appl. Opt.*, 42, 5759, 2003.
23. Yu, P., Peng, L., Mustata, M., Turek, J. J., Melloch, M. R., and Nolte, D. D. Time dependent speckle in holographic optical coherence imaging and the health of tumor tissue, *Opt. Lett.*, 29, 68, 2004.
24. Braga, R. A., dal Fabbro, I. M., Borem, F. V., Rabelo, G., Arizaga, R., Rabal, H. J., and Trivi, M. Assessment of seed's visibility by laser speckle techniques. *Biosys. Eng.*, 86, 287, 2003.
25. Novikoff, A. R. and Holtzman, E., *Cells and Organelles*, Holt, Rinehart and Wiston, 1970.
26. Forrester, A. T., Photoelectric mixing as a spectroscopy tool, *J. Opt. Soc. Am.*, 51, 253, 1961.

27. Cummins, H. Z. and Swinney H. L. The theory of light beating spectroscopy, in *Progress in Optics VIII,* Wolf, E., Ed., 1970, chap. 3.
28. Abramson, N., *Light in Flight or the Holodiagram*, SPIE Press, Bellingham, 1996.
29. Briers, J. D., Laser Doppler and time-varying speckle: A reconciliation. *J. Opt. Soc. Am. A*, 13, 345, 1996.
30. Tuchin, V. V., Laser and fiber optics in biomedicine, *Laser Phys.*, 3, 767, 1993.
31. Hecht, E. and Zajac, A., *Optics,* Addison-Wesley, Reading, MA, 1974.
32. Chu, B., *Laser Light Scattering: Basic Principles and Practice,* 2nd ed., Academic Press, Boston, 1991.
33. Svanberg, S., *Atomic and Molecular Spectroscopy*, Springer Verlag, Berlin, 1992, chap. 4.
34. Optical Society of America, *Handbook of Optics I*, 2nd ed., Bass, M., Ed., McGraw-Hill 1994.
35. Brillouin, L., Diffusion of light and X-ray by a transparent homogeneous body: The influence of thermal agitation. *Ann. Phys.*, 17, 88, 1922.
36. Berg, R., Laser-based cancer diagnostics—Tissue optics considerations, Ph.D. thesis, Lund Institute of Technology, Lund, Sweden, 1995.
37. Xu, Z., Joenathan, C., and Khorana, B M., Temporal and spatial properties of the timevarying speckles of botanical specimens, *Opt. Eng.,* 34, 1487, 1995.
38. Gausman, H. W., Allen, W. A., and Escobar, D. E., Refractive index of plant cell walls, *Appl. Opt.*, 13, 109, 1974.
39. Bergkvist, A., Biospeckle-based study of the line profile of light scattered in strawberries. Master's thesis, Lund University, Sweden, 1997.
40. Martínez, A., Ortiz, C., Arizaga, R., Rabal, H. J., and Trivi, M. Temporal evolution of speckle in foams, *Proceedings of SPIE*, Marcano, A., Ed., 5622, 1484, 2004.
41. Arizaga, R., Cap, N., Rabal, H. J.,Trivi, M., and Baldwin G. Dynamic speckle segmentation, in *Proceeding of the International Symposium on Laser Metrology, "Laser Metrology'99,"* Albertazzi, A., Ed., 1999, 5.73.
42. Drain, L. E., *The Laser Doppler Technique*, John Wiley, New York, 1980, chap. 4.
43. Briers, J. D., Wavelength dependence of intensity fluctuations in laser speckle patterns from biological specimens, *Opt. Comm.* 13, 324, 1975.
44. Gottlieb, C., Bygdeman, M., Thyberg, P., Hellman B., and Rigler, R. Dynamic laser light scattering compared with video micrography for analysis of sperm velocity and sperm head rotation, *Andrology*, 23, 1, 1991.
45. Madigan, M. T., Martinko, J., and Parker, J., *Brock Biology of Microorganisms*, 10th ed., Prentice Hall, Upper Saddle River, NJ, 2002.
46. Zheng, B., Pleass, C. M., and Ih, C. S., Feature information extraction from dynamic biospeckle, *Appl. Opt.,* 33, 231, 1994.
47. Gaskill, J. D., *Linear System, Fourier Transforms, and Optics*, John Wiley New York, 1978.
48. Osten, W. Baumbach, T., Seebecher, S., and Juptner, W. *Remote Shape Control by Comparative Digital Holography,* Fringe 2001, Osten, W., Juptner, W., Ed., Elsevier, 2001, 373–382.
49. See Lee and Harp model in Frieden, B. R., *Probability, Statistical Optics, and Data Testing,* Springer-Verlag, Berlin, 1983, pp. 92.
50. Armstrong, B. H., Spectrum line profiles: The Voigt profile, *J. Quan. Spectr. Rad. Trans.*, 7, 61, 1967.
51. Duncan, D. D. and Kirkpatrick, S. J., The copula: A tool for simulating speckle dynamics, *J. Opt. Soc. Am. A*, 25, 231, 2008.

3 Speckle and Dynamic Speckle Phenomena

Theoretical and Numerical Models

María Fernanda Ruiz Gale, Elsa Noemí Hogert,
Néstor Gustavo Gaggioli, Hector Jorge Rabal,
and Antônio Tavares da Costa Júnior

CONTENTS

3.1 INTRODUCTION

The body boundary that separates and differentiates it from the outside is usually called *surface*. Technically speaking, the surface is the interface between two media with different properties. Its shape can offer information about some background processes. When the surface has been machined, polished, worn, eroded, milled, extruded, etc., a characteristic structure, known as *texture*, normally appears on it.[1,2] The topographical description and statistical properties of this texture are the purpose of many research works.

Material anomalies due to erosion have often led to unforeseen failures in many mechanical parts. Several classical techniques are commonly used to address the problem,[3] such as x-radiography, thermal neutron radiography, ultrasonic testing, eddy current testing, acoustic emission testing, etc. Some optical methods are currently applied to detect impact damage and delamination of composites,[3–10] but, generally speaking, the quest for a rapid, sensitive, and automatic nondestructive testing (NDT) technique still remains unsatisfied.

Furthermore, there are many cases when it is important to detect changes on rough surfaces using nondestructive and noninvasive techniques. Such surfaces can be dielectric, metallic or organic, and hard or soft. Besides, the changes can be produced by oxidation, erosion, degradation, growth in unsprayed samples, bloody flow, evaporation of liquids, etc.

The experience achieved after the first works on speckle in the early seventies, and later studies of rough surfaces by means of light scattering are an excellent background for theoretical and numerical models to study small changes in rough surfaces through the analysis of first- and second-order statistical properties of scattered light. On the other hand, some current requirements of NDT may be solved by a study of the variations in the scattered light due to small changes in a rough surface as a way to predict the process.

Therefore, main interest is to develop theoretical and numerical models to predict the erosion evolution by analyzing the changes in the scattered light, which can result in a NDT method.[11–17,24] First- and second-order statistical properties of light scattered in the far-field region are frequently used for measuring surface characteristics.[11]

3.2 THEORETICAL MODEL

3.2.1 Variable Surface Model

In many cases the roughness changes in dielectric or metallic surfaces are important and need to be detected by means of an NDT method. Therefore, the study of light scattered from changing surfaces (oxidation, erosion, biological samples, etc.), is of great interest because of its applications in different areas such as aging control of mechanical components, botanical studies, or paint-drying studies.

When a coherent beam illuminates a rough surface, the scattered light has a speckle pattern. It is well known that the structure of this pattern depends on surface texture and illumination conditions. Therefore, any change in these conditions produces related changes in the speckle pattern. This behavior has been widely studied since the first work of Léger.[18] Two speckle patterns, before and after any change occurs, are correlated to analyze this phenomenon.

A frequent application of the speckle correlation method is the study of surface roughness.[19–24] In this case, only the incident angle is varied so that the angular speckle correlation obtained depends just on surface roughness and angular variation. This is an example of a stationary speckle pattern.

A dynamic speckle pattern occurs when an illuminated surface presents any kind of activity. Particle movement, any kind of morphological changes of the surface (biological samples, paint drying), refractive index variation, etc., produce a dynamic speckle.[23–30] Many industries are interested in the study of such surfaces, but until laser speckle techniques became available, the usual methods were visual or gravimetric.

In this section, a quasidynamic model for the dynamic speckle is described. In this case, there are no changes in illumination conditions; only the surface morphology is varied for transmission or reflection geometry in a controlled way. A theoretical expression relating the speckle intensity correlation coefficient as a function of surface roughness variation is shown. The different cases in which a surface changes are referred to as "changing surfaces."

In order to analyze the quasidynamic model developed in this chapter, it is important to know the surface roughness variation corresponding to each speckle pattern. Due to the difficulty in evaluating the instantaneous roughness of a real changing surface, a system that behaves similar to a surface with a roughness that could be controlled is used. A digital speckle correlation (DSC) coefficient as a function of roughness changes is determined and compared with the model.

3.2.2 ROUGHNESS OF A VARIABLE SURFACE

Let the function $\zeta(x, y)$ be the height of the surface at the point (x, y). It is assumed that the initial height undergoes a perturbation $\delta\zeta(x, y)$ such that the height of the new surface is the old one plus the perturbation, that is, $\zeta'(x, y) = \zeta(x, y) + \delta\zeta(x, y)$. It can be considered that both ζ and $\delta\zeta$ are random functions, with Gaussian probability density function (pdf). Perturbation ζ has variance σ^2 and mean value 0.

The variance of $\delta\zeta$ is $\delta\sigma^2 = <(\delta\zeta - <\delta\zeta>)^2>$, and its mean value is μ. Let the standard deviations σ and $\delta\sigma$ be defined as the surface "roughness" and "roughness variation," respectively.

If ζ and $\delta\zeta$ are jointly normal, their joint pdf is[31–33]

$$f_{\zeta,\delta\zeta}(\zeta,\delta\zeta) = \frac{1}{2\pi\,\sigma\,\delta\sigma\sqrt{1-r^2}}\exp\left\{-\left[\frac{1}{2(1-r^2)}\right]\left[\frac{(\zeta)^2}{\sigma^2} - \frac{2r\zeta\delta\zeta}{\sigma\,\delta\sigma} + \frac{(\delta\zeta-\mu)^2}{\delta\sigma^2}\right]\right\}$$

(3.1)

The parameter r is the correlation coefficient between both random variables. It can be easily seen that the pdf of ζ and $\delta\zeta$ are also normal, and the pdf of the sum of these two random variables is

$$f(\zeta') = \frac{1}{\sqrt{2\pi}\sigma'}\exp\left[-\frac{(\zeta'-\mu)^2}{2\sigma'^2}\right]$$

(3.2)

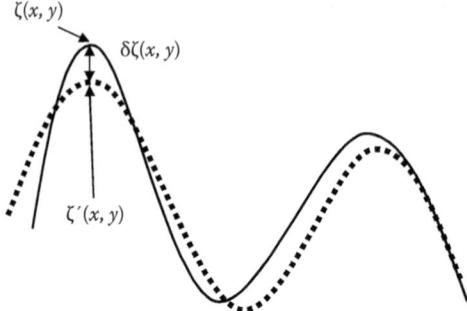

FIGURE 3.1 Model of the variable surface profile.

where $\sigma' = \sqrt{\sigma^2 + 2r\sigma\,\delta\sigma + \delta\sigma^2} = \sqrt{\langle(\zeta' - <\zeta>)^2\rangle}$ is the roughness of the new surface, and $\langle\zeta'\rangle = \langle\delta\zeta\rangle = \mu$ is its mean value.

Two limit cases can be considered:

a. The random variables are independent, $r = 0$; the roughness is $\sigma' = \sqrt{\sigma^2 + \delta\sigma^2}$.
b. $\delta\zeta$ is a function of ζ, $r = \pm1$, and $\sigma' = \sigma \pm \delta\sigma$.

The experimental setup described in 3.2.2.4 satisfies the second case because the surface profile is changed in a monotonic way with an immersion liquid in contact with the rough surface. Figure 3.1 presents the models of the variable surface profile.

3.2.2.1 Speckle Intensity Correlation

Speckle intensity correlation considers a surface illuminated with a coherent light and the associated speckle pattern formed in the scattered field. We will observe only objective speckle—the random elementary wave phases propagating freely in space, without the imaging system. When the surface roughness changes an amount $\delta\sigma$, the scattered speckle pattern also changes. This variation can be quantified by the speckle intensity correlation coefficient ρ_I that depends on surface roughness $\delta\sigma$, the incidence angle θ_1, and the observation angles θ_2 and θ_3 (See Figure 3.2).

In order to find an expression for the correlation coefficient, it is assumed that the speckle pattern is fully developed, and the corresponding complex amplitude $E(\sigma, \theta_1, \theta_2, \theta_3)$ can be treated as a circular complex random Gaussian process.[31,33] Then, the speckle intensity correlation ρ_I between two speckle patterns, before and after the roughness surface changes without changing the incidence angle, and can be evaluated from

$$\rho_I = \frac{|\langle E(\sigma, \theta_1, \theta_2, \theta_3) E'^*(\sigma', \theta_1, \theta_2, \theta_3)\rangle|^2}{\langle I(\sigma, \theta_1, \theta_2, \theta_3)\rangle\langle I'(\sigma, \theta_1, \theta_2, \theta_3)\rangle} \tag{3.3}$$

where: $\langle I(\sigma, \theta_1, \theta_2, \theta_3)\rangle = \langle E(\sigma, \theta_1, \theta_2, \theta_3) E^*(\sigma, \theta_1, \theta_2, \theta_3)\rangle$ is the mean intensity scattered from the surface with a roughness σ. The prime denotes the variables and functions of the surface after the change, and the asterisk means the complex conjugate.

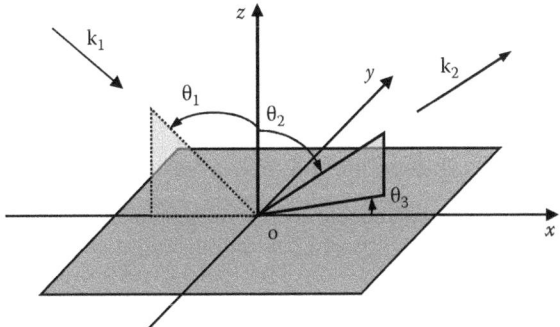

FIGURE 3.2 The basic notation for a reflecting geometry.

In order to evaluate ρ_l, the general Kirchhoff solution for scattering from rough surfaces is used.

$$E(\theta_1,\theta_2,\theta_3) = K_{2R} \int_{-L}^{L} \exp[jvr]\,dx\,dy \tag{3.4}$$

where $v = (v_x, v_y, v_z) = k_1 - k_2$; k_1 and k_2 are the incident and observation wave vectors, respectively; $r = (x, y, \zeta(x, y))$ is the coordinate of a surface point; and K_{2R} is a geometric factor.

$$K_{2R} = Qk\,\frac{1 + \cos\theta_1\cos\theta_2 - \mathrm{sen}\,\theta_1\mathrm{sen}\,\theta_2\cos\theta_3}{\cos\theta_1 + \cos\theta_2} \tag{3.5}$$

where $k = |k|$, and when the surface is smooth,[15–30,32] the geometric factor K_{2R} results in an imaginary number proportional to the transmission or reflexion coefficient.

After the change, the scattered field has an analogous expression similar to Equation 3.4 but using K'_{2R}, v', and r' in the corresponding places. Because there are no changes in the illumination conditions, it results in $|K_{2R}| = |K'^*_{2R}|$. The numerator of Equation 3.3 becomes

$$|\langle E\ E'^* \rangle|^2 =$$

$$\left| \left\langle |K_{2R}|^2 \int_{-L}^{L} \exp[j(v_x x_1 + v_y y_1 + v_z \zeta_1)]\,dx_1\,dy_1 \int_{-L}^{L} \exp[-j(v_x x_2 + v_y y_2 + v_z \zeta_2')]\,dx_2\,dy_2 \right\rangle \right|^2 \tag{3.6}$$

where $\zeta_1 = \zeta(x_1, y_1)$, $\zeta_2 = \zeta(x_1, y_1)$, and L is the side of the square illuminated area.

As the incidence and observation angles are not changed, $v' = v$, and then

$$|\langle E\ E'^* \rangle|^2 = |K_{2R}|^4 \left| \int_{-L}^{L}\int_{-L}^{L} \left\langle e^{jv_z(\zeta_1 - \zeta_2')} \right\rangle e^{jv_x(x_1 - x_2) + jv_y(y_1 - y_2)}\,dx_1\,dx_2 \right|^2 \tag{3.7}$$

In order to evaluate this integral, two cases have been considered: (1) $\delta\zeta(x, y)$ and $\zeta(x, y)$ are independent random variables, and (2) $\delta\zeta(x, y)$ and $\zeta(x, y)$ are dependent random variables. After that, the speckle intensity correlation coefficient for reflection and transmission geometries can be calculated.

3.2.2.2 $\delta\zeta(x)$ and $\zeta(x)$ Independent Random Variables

If $\delta\zeta(x, y)$ and $\zeta(x, y)$ are independent random variables, the mean value of the inner integral of Equation 3.7 can be expressed as a product of two mean values:

$$\langle \exp[jv_z(\zeta_1 - \zeta_2')] \rangle = \langle \exp[jv_z(\zeta_1 - \zeta_2)] \rangle \langle \exp[-jv_z\delta\zeta_2] \rangle = \chi_2(v_z, -v_z)\,\chi_1(-v_z) \quad (3.8)$$

where $\chi_1(-v_z) = \langle \exp[-jv_z\delta\zeta] \rangle$ is the characteristic function associated with the random variable $\delta\zeta$, and $\chi_2(v_z, -v_z) = \langle \exp[jv_z(\zeta_1 - \zeta_2)] \rangle$ is the two-dimensional characteristic function of the random variables ζ_1 and ζ_2, which are related by the autocorrelation coefficient of the surface $C(\tau)$, being τ the distance between any two points (x_1, y_1) and (x_2, y_2).

As ζ_1 and ζ_2 are jointly normal, and $\delta\zeta$ is also normal, Equation 3.8 becomes

$$\chi_2(v_z, -v_z)\,\chi_1(-v_z) = \exp\left[-v_z^2\sigma^2(1 - C(x_1 - x_2))\right]\exp\left[-\frac{v_z^2\delta\sigma^2}{2}\right]\exp[-iv_z\mu] \quad (3.9)$$

Substituting Equation 3.9 in Equation 3.8 and this in turn in Equation 3.7, making the variable changes $x_1 - x_2 = \tau$; $x_2 = x$, and integrating on x results in

$$\left|\langle E(\theta_1, \theta_2)E'^*(\theta_1, \theta_2)\rangle\right|^2$$

$$= |K_{2R}|^4 \left|\exp\left[-\frac{v_z^2\delta\sigma^2}{2}\right]\exp[-iv_z\mu]\,2L\int_{-L}^{L}\exp\left[-v_z^2\sigma^2(1 - C(\tau))\right]\exp[jv_x\tau]\,d\tau\right|^2$$

$$(3.10)$$

Usually, $C(\tau)$ can be considered as a Gaussian function: $C(\tau) = \exp[-\tau^2/T^2]$, where T is the "correlation distance."

Considering a very rough surface, that is,. $\lambda \ll \sigma$, it can be easily verified that the only significant contribution to integral (Equation 3.10) is that from the region near to $\tau = 0$. Thus, setting $\exp(-\tau^2/T^2) \approx 1 - \tau^2/T^2$ and replacing the limits of integration by $\pm\infty$, we have

$$\left|\langle E(\theta_1, \theta_2)E'^*(\theta_1, \theta_2)\rangle\right|^2 = 4L^2|K_{2R}|^4\exp\left[-v_z^2\delta\sigma^2\right]\left|\int_{-\infty}^{\infty}\exp\left[-v_z^2\sigma^2\frac{\tau^2}{T^2}\right]\exp[jv_x\tau]\,d\tau\right|^2$$

$$(3.11)$$

Then, integrating Equation 3.11, the numerator of Equation 3.3 results in

$$|\langle E(\theta_1,\theta_2)E'^*(\theta_1,\theta_2)\rangle|^2 = 4\,L^2\,|K_{2R}|^4\,\exp\left[-v_z^2\delta\sigma^2\right]\left(\frac{\sqrt{\pi}\,T}{\sigma v_z}\right)^2\left(\exp\left[-\frac{v_x^2\,T^2}{4\sigma^2 v_z^2}\right]\right)^2$$

(3.12)

In order to evaluate the denominator of Equation 3.3, the mean scattered intensity, as found by Beckmann, is considered with the modifications for the experimental changing surface model. Equation 3.3 can be written as

$$\rho_I = \frac{\dfrac{T}{\sigma}\exp\left(-v_z^2\delta\sigma^2\right)\exp\left(-\dfrac{v_x^2\,T^2}{4\,v_z^2\sigma^2}\right)}{\dfrac{T'}{\sigma'}\exp\left(-\dfrac{v_x^2\,T'^2}{4\,v_z^2\sigma'^2}\right)}$$

(3.13)

For the specular direction $v_x = 0$, v_z vanishes only for grazing incidence and reflection, that is, $\theta_2 = \theta_1 \cong \pi/2$. In the former case $\theta_1 = \theta_2 = \theta$, the speckle intensity correlation is reduced to

$$\rho_I = \frac{\dfrac{T}{\sigma}\exp\left[-v_z^2(\theta)\delta\sigma^2\right]}{\dfrac{T'}{\sigma'}}$$

(3.14)

v_z is different for reflecting and refracting geometry.

At this point, let us note that the model for changing surfaces was developed for studying the beginning of these processes. Therefore, roughness and correlation distance changes will be very small. This means that

$$\delta\sigma \ll \sigma$$

(3.15)

$$T' = T + \delta T \quad \text{and} \quad \delta T \ll T$$

(3.16)

Moreover, expression 3.14 may be considered as the product of two terms. The negative exponential has no relevant values for $\rho_I < 0.1$ (experimental limitation). For this reason, $\delta\sigma$ must be smaller than 0.2 µm, and thus it is reasonable to consider $\delta\sigma \ll \sigma$. The other term of Equation 3.14, $T\sigma'/T'\sigma$, may be approximated to a straight line $(1 - \delta T/T)$ and, for the case in point, it could be approximated by 1.

Finally, the speckle intensity correlation coefficient will be

$$\rho_I = \exp\left(-v_z^2(\theta)\,\delta\sigma^2\right)$$

(3.17)

3.2.2.3 $\delta\zeta(x)$ and $\zeta(x)$ Dependent Random Variables

In order to obtain a controlled changing roughness $\delta\sigma$, an experimental setup in which the rough surface is immersed in a liquid of refractive index n_2 was designed. The scattered light behaves as if the incident beam had crossed another surface whose roughness is termed *apparent*. In this case, $\delta\zeta$ is a linear function of ζ; therefore, the mean value of integral 3.7 must be evaluated differently. If $\delta\zeta = \alpha\zeta$, the height of the new surface becomes $\zeta' = \zeta + \delta\zeta = \zeta + \alpha\zeta = \beta\zeta$, and $\sigma' = \beta\sigma$.

The inner integral mean value (Equation 3.7) is the two-dimensional characteristic function $\chi(v_z, - \beta v_z)$ of the random variables ζ_1 and ζ_2:

$$\langle \exp[j(v_z\zeta_1 - v_z\zeta_2')]\rangle = \langle \exp[j[v_z\zeta_1 - v_z\beta\zeta_2]]\rangle = \chi(v_z, -\beta v_z) \tag{3.18}$$

The characteristic function (Equation 3.18) is similar to that obtained in the angular speckle correlation method.[12–19] In that,

$$\chi(v_z, -\beta v_z) = \exp\left[-\frac{\sigma^2}{2}(v_z^2 + \beta^2 v_z^2 - 2\beta\, C(\tau)\, v_z^2)\right] \tag{3.19}$$

where the earlier definition for τ has been used. Substituting Equation 3.19 in Equation 3.18 and this in turn in Equation 3.7, making the same variable changes, and integrating in x results in

$$\left|\langle E(\theta_1, \theta_2) E'^*(\theta_2, \theta_2)\rangle\right|^2$$

$$= 4L^2 |K_{2R}|^4 \exp[-\sigma^2 v_z^2(1-\beta)^2] \left|\int_{-L}^{L} \exp[-\sigma^2\, v_z^2\, \beta(1-C(\tau))] \exp[jv_x\tau]\,d\tau\right|^2 \tag{3.20}$$

As before, we can do the approximation $\exp(-\tau^2/T^2) \approx 1 - \tau^2/T^2$ and replace the limits of integration by $\pm\infty$. After integrating,

$$\left|\langle E(\theta_1, \theta_2) E'^*(\theta_1, \theta_2)\rangle\right|^2 = 4L^2 |K_{2R}|^4 \exp[-\sigma^2 v_z^2(1-\beta)^2] \frac{\pi T^2}{\beta\cdot\sigma^2 v_z^2} \exp\left[-\frac{v_x^2 T^2}{2\beta\,\sigma^2 v_z^2}\right] \tag{3.21}$$

Substituting Equation 3.21 and the mean scattered intensities in Equation 3.3 and considering $\theta_1 = \theta_2 = \theta$,

$$\rho_I = \exp[-\sigma^2 v_z^2(1-\beta)^2] = \exp[-v_z^2(\theta)\delta\sigma^2] \tag{3.22}$$

which is the same result of Equation 3.16. Thus, in the present experimental conditions, the same speckle correlation coefficient for the roughness variation is obtained in both cases: $\delta\zeta(x)$ and $\zeta(x)$ dependent or independent random variables.

3.2.2.4 Speckle Intensity Correlation for Reflection and Transmission Geometries

As shown, the correlation coefficient forms are similar. Now, this expression will be analyzed for reflection and transmission geometries because, in these cases, the v_z component changes.

For the reflecting geometry case (Figure 3.3a), $v_z = -kn_0(\cos\theta_1 + \cos\theta_2)$, and Equation 3.17 can be written, when $\theta_1 = \theta$, as

$$\rho_I = \exp[-(2kn_0\cos\theta)^2\,\delta\sigma^2]\tag{3.23}$$

For the transmission geometry case (Figure 3.3b),

$$v_z = k(n\cos\theta_t - n_0\cos\theta_2) = k(\sqrt{n^2 - n_0^2 sen^2\theta_1} - n_0\cos\theta_2)$$

and Equation 3.17 can be written as

$$\rho_I = \exp\left\{-\left[k^2\left(\sqrt{n^2 - n_0^2 sen^2\theta} - n_0\cos\theta\right)^2\delta\sigma^2\right]\right\}\tag{3.24}$$

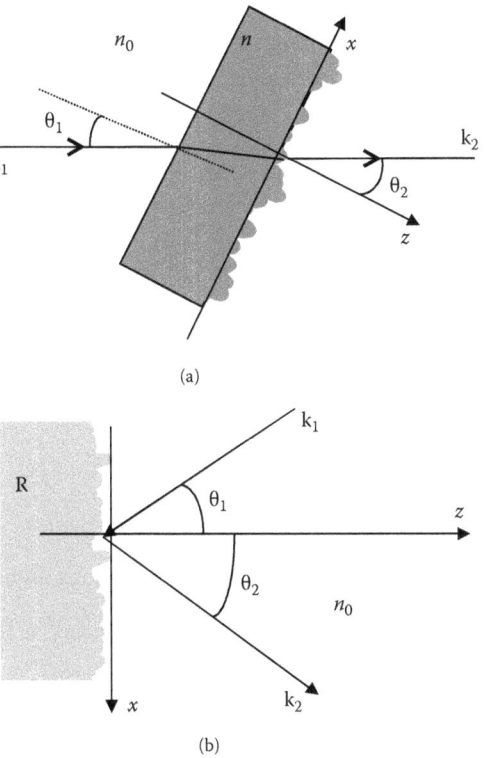

(a)

(b)

FIGURE 3.3 Basic notation for (a) transmission geometry and (b) reflecting geometry.

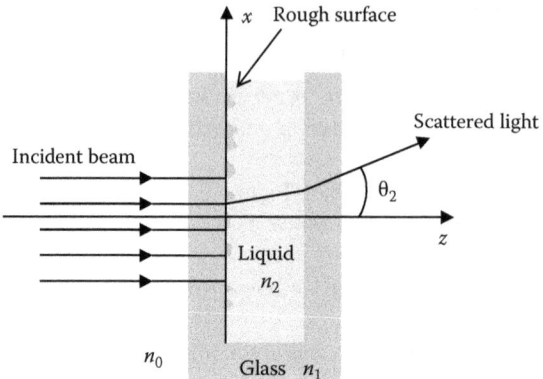

FIGURE 3.4 Schematic view of the experimental geometry cell.

The actual model allows the study of random surfaces with changing roughness in the reflection or transmission mode. Nevertheless, for simplicity, the transmission geometry was used. A controlled changing roughness can be simulated using the cell described in the following text and is illustrated in Figure 3.4. The rough surface under study is the inner surface of a glass cell with refractive index $n = 1.52$ and a Gaussian statistic. The cell is filled with liquids having different refractive indices n_2. Solutions of acetone ($n_{ac} = 1.3556$) with small additions of CS_2 ($n_{CS_2} = 1.6204$) were used. It is thus possible to simulate different roughness.

In the experimental geometry, $\theta_1 = \theta_2 = 0$, and the cell is immersed in air, then $n_0 = 1$. The roughness of a cell–liquid system (Figure 3.4) viewed from the air in the positive z-axis is named *apparent roughness*. This can be written as

$$\sigma_a = \sigma\frac{n - n_2}{n - 1} = \sigma - \sigma\frac{n_2 - 1}{n - 1} \quad (n \neq 1) \tag{3.25}$$

Therefore, modifying the refractive index of the immersion liquid, the apparent roughness changes, and the roughness variation is

$$\delta\sigma_a = -\frac{\sigma}{n - 1}\delta n_2 \tag{3.26}$$

On the other hand, for the present experimental conditions, Equation 3.24 becomes

$$\rho_I = \exp\left[-k^2(n - 1)^2\delta\sigma_a^2\right] \tag{3.27}$$

Substituting Equation 3.26 in Equation 3.27, the correlation coefficient for transmission geometry is

$$\rho_I = \exp\left[-k^2\sigma^2\delta n_2^2\right] \tag{3.28}$$

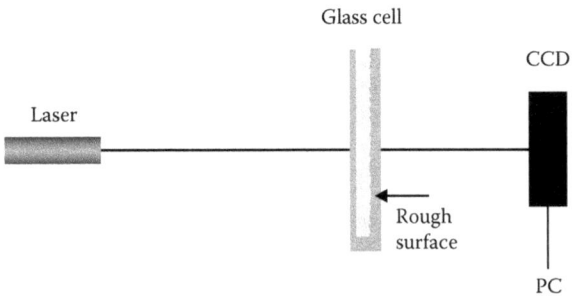

FIGURE 3.5 Experimental setup for roughness investigation.

This result coincides with that obtained in a previous work[16] for a different approach and application.

Figure 3.5 shows the experimental setup. After traversing a pupil (circular aperture ≈ 4 mm²), a collimated He–Ne laser beam (λ = 0.633 μm) impinges perpendicularly on the plane face of a ground glass from air. A charge-coupled device (CCD) camera placed on the observation plane records and stores the speckle intensity patterns. The size of the illuminated area and speckle pattern must satisfy the Shannon ergodic theorems.[31] The speckle patterns produced by different refractive indices (i.e., different roughness variations) are stored in the personal computer (PC). Subsequently, a mathematical algorithm evaluates the digital correlation between the first speckle pattern and the following ones. The digital correlation coefficient in the function of the $\delta\sigma$ value is obtained. In this way, it is possible to verify the theoretical model. Two ground glasses were used with σ_1 = 3.5 μm, σ_2 = 1.2 μm, and T_1 = 45 μm, T_2 = 20 μm.

Using these experiments, the intensity correlation coefficient ρ_I as a function of roughness variations $\delta\sigma$ is obtained. The correlation coefficient ρ_I versus $\delta\sigma$ for different initial roughness is plotted (Figure 3.6). It can be noticed that ρ_I versus. $\delta\sigma$

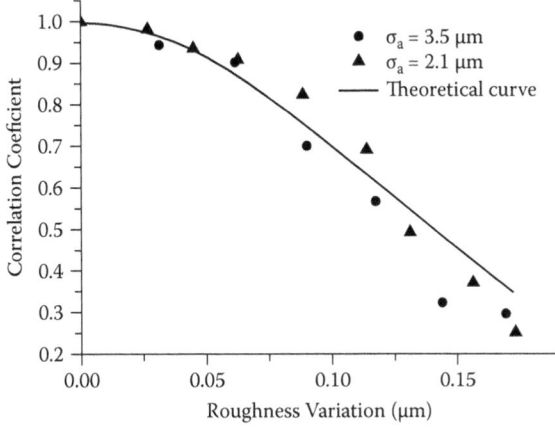

FIGURE 3.6 Intensity correlation coefficient versus roughness variations.

is well fitted by a decreasing quadratic exponential function, as it was predicted in Equations 3.17 and 3.27.

The result shows that this model can be used to analyze rough surfaces undergoing a process that modifies their roughness.

3.2.2.5 Example: Wet Paint Drying on a Rough Surface

Rough surfaces covered with a wet synthetic paint layer were studied using the speckle correlation method. The rough surface was a bronze sample with an initial roughness $\sigma_0 = 7.1$ μm. A thin layer of synthetic paint was placed over it and, after a couple of minutes, the measurement was started.

Figure 3.7 shows the experimental device. A collimated He–Ne laser beam ($\lambda = 0.633$ μm, area ~1 mm²) impinges perpendicularly on the painted rough surface. A CCD camera placed near the normal direction ($\varphi \sim 0{,}07$ rad) records the speckle intensity patterns. A set of speckle images produced by different roughness variations due to the paint drying is displayed on the screen and stored in a PC. The digital correlation coefficient between the first speckle pattern and the following ones as the function of time is obtained.

When the incident and observation angles are equal to zero ($\theta = \theta_1 = 0$), the correlation coefficient from Equation 3.24 is

$$\rho_I = \exp\left[-(2k)^2\,\delta\sigma^2\right] \tag{3.29}$$

The roughness variation is thus

$$|\delta\sigma| = \frac{\sqrt{-\ln\rho_I}}{2k} \tag{3.30}$$

Figure 3.8a and b shows the time variation of the digital speckle correlation and the roughness variation.

Because the solvent concentration as a function of t is likely to be a decreasing exponential function,[7,8] it is assumed that the fitting curve for the roughness variation is also a decreasing exponential function as

$$\delta\sigma(t) = \delta\sigma_0\left[1 - \exp\left(-\frac{t}{T_\rho}\right)\right] \tag{3.31}$$

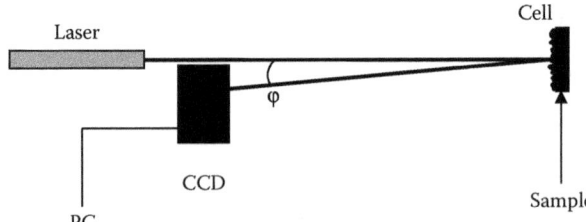

FIGURE 3.7 Experimental setup to analyze paint drying in a rough surface.

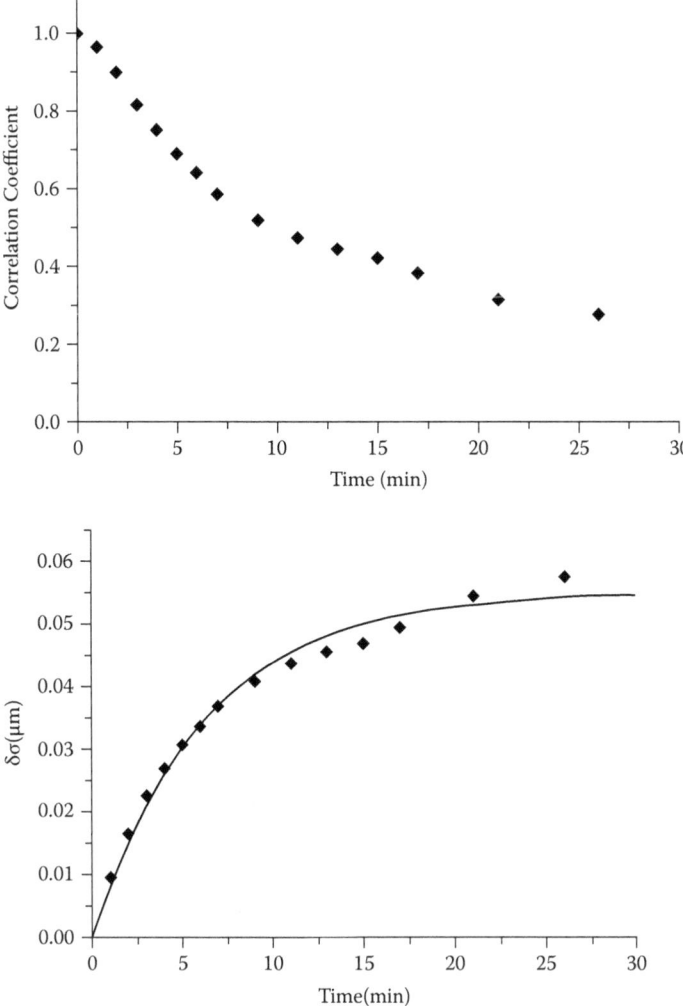

FIGURE 3.8 Curves of (a) speckle correlation and (b) roughness variation.

Thus, after fitting with a exponential expression (Equation 3.31), the following parameters are obtained:

$$\delta\sigma_0 = (0,055\pm0,001)\mu m \text{ and } T_\rho = (6,25\pm0,4) \text{ min}$$

and are related to $\delta\sigma$ random processes. Its relative error is less than 10%.

Because the surface activity is high 30 min after paint application, the speckle correlation value is equal to the measurement error though the paint is still wet.

In order to study the long-time behavior, a different test was therefore needed. A new thin layer of synthetic paint was placed over another portion of the rough surface, sample B, and a set of speckle patterns was recorded every 20 min. The speckle correlation was

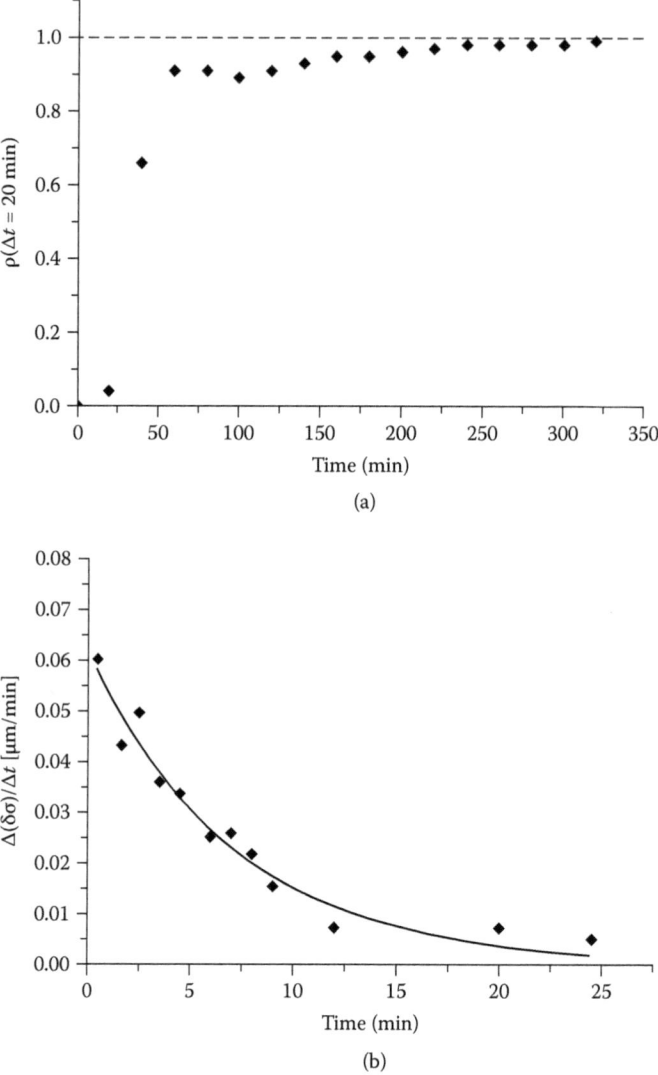

FIGURE 3.9 Speckle correlation (a) between consecutive images taken every 20 min, and (b) roughness change rate versus time.

performed between consecutive speckle patterns (Figure 3.9a), that is, it was evaluated between the first and the second picture, the second and third picture, and so on.

The first point in Figure 3.9a corresponding to the first 20 min shows a high activity because the correlation is zero. This time period was studied in the previous paragraph. The second point $\rho = 0.7$ indicates that the surface has a decreasing activity. After 60 min, the speckle correlation is larger than 0.9, as shown by the third point. After that, the correlation increases slowly, that is, $\rho = 0.99$, 4 h and 40 min later. From Equation 3.30, this correlation value corresponds to a roughness variation $\sigma_\rho = 5$ nm for the last 20 min. This is one possibility among many criteria for "activity degree" that can be used depending on the requirement of the case.

Several authors have carried out empirical analysis of speckle dynamic correlation. Nevertheless, the present approach makes it possible to quantify the surface activity through Equation 3.31. The roughness change rate is defined as

$$v_a = \frac{\delta\sigma_{m,m+1}}{\delta t_{m,m+1}} \tag{3.32}$$

where $\delta t_{m,m+1} = t_{m+1} - t_m$ is the time interval between m and $m+1$ states

$$\delta\sigma_{m,m+1} = \frac{\sqrt{-\ln\rho_{m,m+1}}}{2k} \tag{3.33}$$

and $\rho_{m,m+1}$ is the correlation coefficient between m state and $m+1$ state.

Using the experimental result of Figure 3.9a, where $\delta t_{m,m+1} = 1$ min, V_a versus t_m is plotted in Figure 3.9b and fit with a decreasing exponential function

$$v_a^{fitted} = B\exp\left(-\frac{t}{T_0}\right) \tag{3.34}$$

with parameters

$$B = (0{,}063 \pm 0{,}003)\mu\text{m/mm} \text{ and } T_0 = (7{,}1 \pm 0{,}5)\text{ min}$$

If $\delta t_{m,m+1}$ is small enough, the time integration of Equation 3.32 yields

$$\delta\sigma(t) = \frac{B}{T_0}\left[1 - \exp\left(-\frac{t}{T_0}\right)\right] \tag{3.35}$$

The preceding equation is similar to Equation 3.31.

3.2.3 ERODED SURFACE MODEL

The aim of this work is to study a particular case of the corrosion process—the electroerosion phenomenon in metallic samples—which is a little bit similar to pitting corrosion. In order to have different erosion degrees, the samples are placed under the same current during several time intervals in an electroerosion machine. Micrographs of some studied samples are presented in Figure 3.10.

Brass and aluminum samples with 1–100% of electroeroded surface were studied, that is, from 1% of the surface covered with pits of electroerosion (this roughness is similar to that of the noneroded surface [$\sigma \cong 1\ \mu$m]) up to 100% erosion ($\sigma \cong 7\ \mu$m).

The optical method used for analyzing this phenomenon assumes a Gaussian statistical distribution of the surface heights.[7,9–11] However, when the surface is partially corroded, the relief departs from a Gaussian distribution. Therefore, some corrections to the original model are necessary. As a first approach, the roughness with increase of corroded surface will be studied by means of a simple model.

3.2.3.1 Roughness of an Electroeroded Surface

In this model, it has been assumed that the noneroded surface has a height Gaussian distribution $N_1(\sigma_1, 0)$, with a roughness σ_1 and mean value 0. When the surface is attacked by an electroerosion process, pits of $2h$ depth with bottom roughness σ_2 and

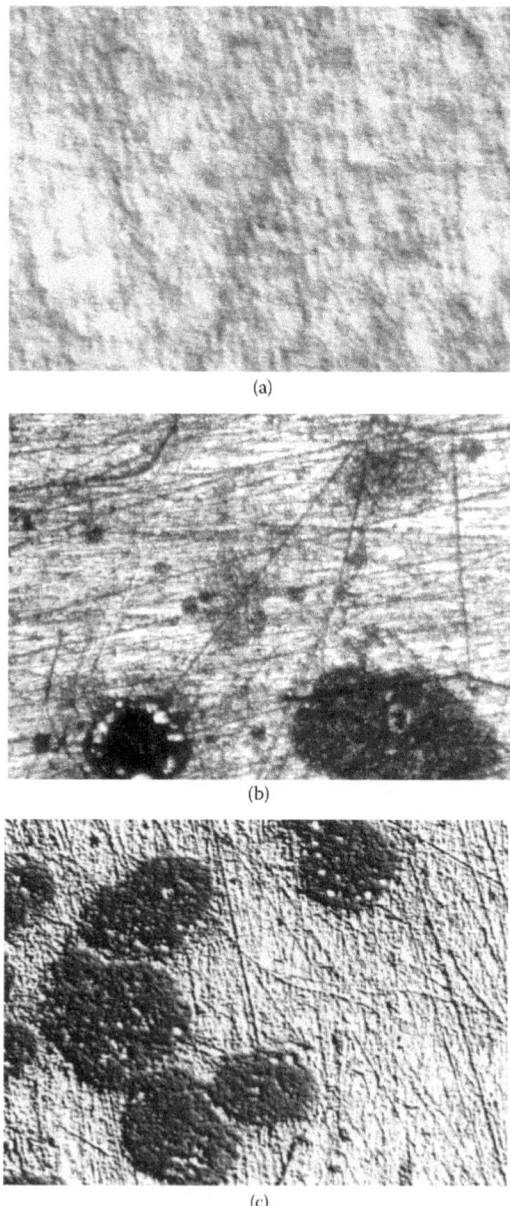

(a)

(b)

(c)

FIGURE 3.10 Electroeroded surfaces with different densities of erosion: (a) 1%, (b) 10%, (c) 30%, and (d) 100%.

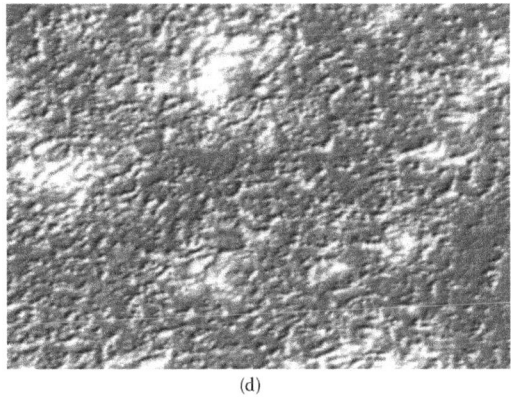

(d)

FIGURE 3.10 (*Continued*)

height Gaussian distribution $N_2(\sigma_2, 0)$ are produced by the spark (Figure 3.11). For simplicity, a unidirectional surface is considered.

Three random variables have been defined to describe $\zeta(x)$:

z_0—related to the pit distribution. It can only take the values h and $-h$.

z_1—related to the nonelectroeroded surface roughness. A Gaussian probability density function $N_1(\sigma_1, 0)$ has been assumed.

z_2—related to the roughness of the pit bottom (nucleus of erosion). A Gaussian probability density function $N_2(\sigma_2, 0)$ has also been assumed.

The coordinate system has been defined such that the height origin corresponds to the middle of the well. Therefore, the mean value of the height is, in general, different from zero. The surface height for any position x on the specimen can be written as

$$\zeta(x) = z_0(x) + z_1(x)\ \delta_{z_0,h} + z_2(x)\ \delta_{z_0,-h} \tag{3.36}$$

where $\delta_{i,j}$ is the Kronecker delta function.

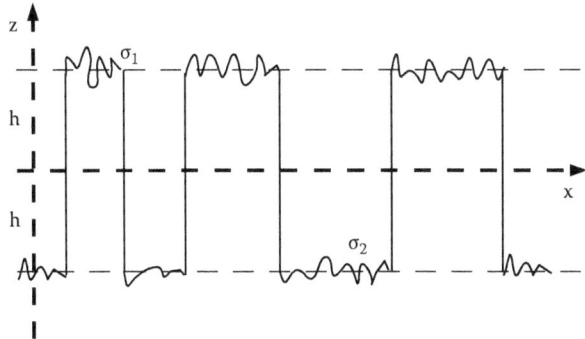

FIGURE 3.11 Model of the electroeroded surface profile.

The ratio of the electroeroded surface to the total surface will be named as s. If the surface is nonelectroeroded, $s = 0$ and if it is completely eroded, $s = 1$, then the probability function for z_0 is

$$p(z_0) = \begin{cases} (1-s) \rightarrow z_0 = h \\ s \rightarrow z_0 = -h \\ 0 \rightarrow z_0 \neq h, -h \end{cases}$$

(3.37)

If the three random variables are independent,

$$p(\zeta) = p(z_0) \, N_1(\sigma_1, 0) \, N_2(\sigma_2, 0)$$

(3.38)

Taking into account that the surface roughness is

$$\sigma = \sqrt{\frac{1}{L} \int_0^L \zeta^2(x) \, dx - \left(\frac{1}{L} \int_0^L \zeta(x) \, dx \right)^2}$$

(3.39)

in order to evaluate the height variance σ^2

$$\sigma^2 = \int_{-\infty}^{\infty} \zeta^2 \, p(\zeta) \, d\zeta - \left(\int_{-\infty}^{\infty} \zeta \, p(\zeta) \, d\zeta \right)^2 = \langle \zeta^2 \rangle - \langle \zeta \rangle^2$$

(3.40)

we first calculate

$$\langle \zeta \rangle = \int \zeta \, p(\zeta) \, d\zeta = \int \left[z_0 + z_1 \, \delta_{z_0, h} + z_2 \, \delta_{z_0, -h} \right] p(z_0) \, N_1(\sigma_1, 0) \, N_2(\sigma_2, 0) \, dz_0 \, dz_1 \, dz_2$$

$$= \langle z_0 \rangle + (1-s) \langle z_1 \rangle + s \langle z_2 \rangle$$

(3.41)

Besides,

$$\langle z_0 \rangle = h \, (1 - 2s)$$

$$\langle z_1 \rangle = 0$$

$$\langle z_2 \rangle = 0$$

Equation 3.41 then becomes

$$\langle \zeta \rangle = h(1 - 2s)$$

(3.42)

The other term of the Equation 3.35 is

$$\langle \zeta^2 \rangle = \int \zeta^2 \, p(\zeta) \, d\zeta = \int \left[z_0 + z_1 \, \delta_{z_0,h} + z_2 \, \delta_{z_0,-h} \right]^2 p(z_0) N_1(\sigma_1,0) N_2(\sigma_2,0) dz_0 dz_1 dz_2$$

(3.43)

After calculation, we have

$$\langle \zeta^2 \rangle = h^2 + (1-s)\sigma_1^2 + s\sigma_2^2$$

(3.44)

As the variance of the variable z_0 is

$$\sigma_0 = 2h\sqrt{s\,(1-s)}$$

(3.45)

from Equations 3.42, 3.44, and 3.45, the total roughness is

$$\sigma^2 = \sigma_0^2 + (1-s)\sigma_1^2 + s\sigma_2^2$$

(3.46)

and in terms of s is

$$\sigma^2 = -4h^2 s^2 + (4h^2 - \sigma_1^2 + \sigma_2^2)s + \sigma_1^2 = -p^2 s^2 + (p^2 - \sigma_1^2 + \sigma_2^2)s + \sigma_1^2$$

(3.47)

where $p = 2h$ is the well depth.

3.2.3.2 Mean Intensity of an Electroeroded Surface

From the general Kirchhoff solution for scattering from a one-dimensional rough surface, the mean intensity of the scattered light (Equation 3.4) results in

$$|\langle E \, E^* \rangle|^2 = |K_{2R}|^4 \left| \int_{-L}^{L} \int_{-L}^{L} \left\langle e^{jv_z (\zeta_1 - \zeta_2)} \right\rangle e^{jv_x (x_1 - x_2)} \, dx_1 dx_2 \right|^2$$

(3.48)

Taking into account that

$$\zeta_1(x_1) = z_0(x_1) + z_1(x_1) \cdot \delta_{z_0,h} + z_2(x_1) \cdot \delta_{z_0,-h}$$

(3.49)

and

$$\zeta_2(x_2) = z_0(x_2) + z_1(x_2) \cdot \delta_{z_0,h} + z_2(x_2) \cdot \delta_{z_0,-h}$$

(3.50)

z_0, z_1, and z_2 are independent but not x_1 and x_2; therefore, the jointly probability can be written as

$$F_2(\zeta_1, \zeta_2) = p[z_0(x_1), z_0(x_2)] \, N_1[z_1(x_1), z_1(x_2)] \, N_2[z_2(x_1), z_2(x_2)]$$

(3.51)

where $N_1[z_1(x_1), z_1(x_2)]$ and $N_2[z_2(x_1), z_2(x_2)]$ are the Gaussian jointly probabilities, and p is

$$p(z_0(x_1), z_0(x_2)) = \begin{cases} (1-s)^2 \rightarrow z_0(x_1) = z_0(x_2) = h \\ s^2 \rightarrow z_0(x_1) = z_0(x_2) = -h \\ (1-s)s \rightarrow z_0(x_1) = h, z_0(x_2) = -h \\ (1-s)s \rightarrow z_0(x_1) = -h, z_0(x_2) = h \end{cases} \tag{3.52}$$

In this case,

$$\langle e^{it_1 z_1 + it_2 z_2} \rangle$$

$$= \int_{-\infty}^{\infty} \int_{-\infty}^{\infty} f(z_0(x_1), z_0(x_2)) \cdot N_1(z_1(x_1), z_1(x_2)) \cdot N_2(z_2(x_1), z_2(x_2)) \cdot e^{jv_z(\zeta_1 - \zeta_2)} d\zeta_1 d\zeta_2 \tag{3.53}$$

Calling

$$z_0(x_1) = z_{0,1} \qquad\qquad\qquad z_0(x_2) = z_{0,2}$$

$$z_1(x_1) = z_{1,1} \qquad\qquad\qquad z_1(x_2) = z_{1,2}$$

$$z_2(x_1) = z_{2,1} \qquad\qquad\qquad z_2(x_2) = z_{2,2}$$

Replacing Equations 3.49, 3.50, and 3.51 in Equation 3.53, integrating in z_0, and operating,

$$\langle e^{it_1 z_1 + it_2 z_2} \rangle = \iint_{\infty} \iint_{\infty} (1-s)^2 N_1(z_{1,1}, z_{1,2}) \cdot N_2(z_{2,1}, z_{2,2}) \cdot e^{jv_z(z_{1,1} - z_{1,2})} dz_{1,1} dz_{1,2} dz_{2,1} dz_{2,2}$$

$$+ \iint_{\infty} \iint_{\infty} s^2 N_1(z_{1,1}, z_{1,2}) \cdot N_2(z_{2,1}, z_{2,2}) \cdot e^{jv_z(z_{2,1} - z_{2,2})} dz_{1,1} dz_{1,2} dz_{2,1} dz_{2,2}$$

$$+ \iint_{\infty} \iint_{\infty} (1-s) \cdot s \cdot N_1(z_{1,1}, z_{1,2}) \cdot N_2(z_{2,1}, z_{2,2}) \cdot e^{jv_z(-2h + z_{2,1} - z_{1,2} + h_1 - h_2)} dz_{1,1} dz_{1,2} dz_{2,1}$$

$$+ \iint_{\infty} \iint_{\infty} (1-s) \cdot s \cdot N_1(z_{1,1}, z_{1,2}) \cdot N_2(z_{2,1}, z_{2,2}) \cdot e^{jv_z(2h + z_{1,1} - z_{2,2})} dz_{1,1} dz_{1,2} dz_{2,1} dz_{2,2} \tag{3.54}$$

As the first two terms are similar, we will only analyze the first one:

$$\iint_{\infty}\iint_{\infty}(1-s)^2 N_1(z_{1,1},z_{1,2})\cdot N_2(z_{2,1},z_{2,2})\cdot e^{jv_z(z_{1,1}-z_{1,2})}dz_{1,1}dz_{1,2}dz_{2,1}dz_{2,2}=$$

$$=(1-s)^2\chi_{2,1}(v_z,-v_z)$$ (3.55)

where $\chi_{2,1}$ is the characteristic function of surface 1. Let us analyze one of the last two terms of Equation 3.54:

$$\int_{-\infty}^{\infty}\int_{-\infty}^{\infty}\int_{-\infty}^{\infty}\int_{-\infty}^{\infty} z_{2,1}(1-s)\cdot s\cdot N_1(z_{1,1},z_{1,2})\cdot N_2(z_{2,1},z_{2,2})\cdot e^{jv_z(2h+z_{1,1}-z_{2,2})}dz_{1,1}dz_{1,2}dz_{2,1}dz_{2,2}$$

$$=\int_{-\infty}^{\infty}\int_{-\infty}^{\infty}(1-s)\cdot s\cdot N_1(z_{1,1})\cdot N_2(z_{2,2})\cdot e^{jv_z(2h+z_{1,1}-z_{2,2})}dz_{1,1}dz_{2,2}$$

$$=(1-s)\cdot s\cdot e^{j2hv_z}e^{-\frac{\sigma_1^2v_z^2}{2}}e^{-\frac{\sigma_2^2v_z^2}{2}}$$ (3.56)

The last term of Equation 3.54 is similar to Equation 3.56 with the negative complex exponential.

Adding them,

$$(1-s)\cdot s\cdot e^{j2hv_z}e^{-\frac{\sigma_1^2v_z^2}{2}}e^{-\frac{\sigma_2^2v_z^2}{2}}+(1-s)\cdot s\cdot e^{-j2hv_z}e^{-\frac{\sigma_1^2v_z^2}{2}}e^{-\frac{\sigma_2^2v_z^2}{2}}=$$

$$=(1-s)\cdot s\cdot e^{-\frac{\sigma_1^2v_z^2}{2}}e^{-\frac{\sigma_2^2v_z^2}{2}}\left(e^{j2hv_z}+e^{-j2hv_z}\right)=$$

$$=2\cdot(1-s)\cdot s\cdot e^{-\frac{\sigma_1^2v_z^2}{2}}e^{-\frac{\sigma_2^2v_z^2}{2}}\cos(2hv_z)$$ (3.57)

Replacing in Equation 3.43,

$$\langle E_2 E_2^*\rangle=(1-s)^2|K_{2R}|^2\int_{-L}^{L}\int_{-L}^{L}e^{jv_x(x_1-x_2)}\chi_{2,1}(v_z,-v_z)\cdot dx_1 dx_2$$

$$+s^2|K_{2R}|^2\int_{-L}^{L}\int_{-L}^{L}e^{jv_x(x_1-x_2)}\chi_{2,2}(v_z,-v_z)\cdot dx_1 dx_2$$

$$+|K_{2R}|^2 2(1-s)\cdot s\cdot e^{-\frac{\sigma_1^2v_z^2}{2}}e^{-\frac{\sigma_2^2v_z^2}{2}}\cos(2hv_z)\int_{-L}^{L}\int_{-L}^{L}e^{jv_x(x_1-x_2)}dx_1 dx_2$$ (3.58)

The first and second terms can be written using the following identities.

$$\langle I_1 \rangle = |K_{2R}|^2 \int_{-L}^{L} \int_{-L}^{L} e^{jv_x(x_1-x_2)} \chi_2^1(v_z,-v_z) \cdot dx_1 dx$$

$$\langle I_2 \rangle = |K_{2R}|^2 \int_{-L}^{L} \int_{-L}^{L} e^{jv_x(x_1-x_2)} \chi_2^2(v_z,-v_z) \cdot dx_1 dx \qquad (3.59)$$

where $\langle I_1 \rangle$ is the mean intensity scattered by a Gaussian surface free from electroerosion, and $\langle I_2 \rangle$ is the mean intensity scattered by a Gaussian surface with the statistics of the pit bottom surface.

Solving,

$$\langle I \rangle = (1-s)^2 \langle I_1 \rangle + s^2 \langle I_2 \rangle + |K_{2R}|^2 8L^2(1-s) \cdot s \cdot e^{-\frac{\sigma_1^2 v_z^2}{2}} e^{-\frac{\sigma_2^2 v_z^2}{2}} \cos(2hv_z) \cdot \text{sinc}^2(v_x L)$$
$$(3.60)$$

In general, for optically rough surfaces and most-often-used geometries, among the three terms, the third term can be neglected. Therefore, the total mean intensity can be considered as composed by the intensity coming from the first two involved surfaces, and it also depends on the quantity of the eroded surface

$$\langle I \rangle = (1-s)^2 \langle I_1 \rangle + s^2 \langle I_2 \rangle \qquad (3.61)$$

3.2.3.3 Speckle Correlation Measurements

A laser beam impinges onto the surface of a sample, and the speckle patterns are obtained by reflection away from it. The experimental setup is shown in Figure 3.12.

In order to change the incident angle of the laser beam, the sample is placed on a goniometer. A linear CCD camera (with 512 detectors) registers the speckle

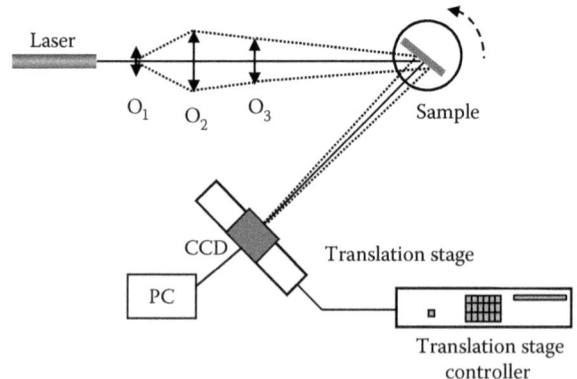

FIGURE 3.12 Experimental setup to analyze speckle correlation measurements. The three lenses O_1, O_2 and O_3 are located such that the laser beam illuminates an adequate area on the sample and in focused on the CCD plane.

intensity on the Fraunhofer plane. In order to follow the speckle pattern, the detector is mounted on a translation stage. The successive unidimensional intensity speckle patterns are recorded in a PC, which is used to process the data.

The correlation ρ_I of the successive speckle patterns, Equation 3.3, is then evaluated by means of a digital mathematical algorithm.

$$\rho_I = \frac{\dfrac{1}{n}\displaystyle\sum_{j=1}^{n}[I_{0j}-\langle I_0\rangle][I_{1j}-\langle I_1\rangle]}{\left[\dfrac{1}{n}\displaystyle\sum_{j=1}^{n}[I_{0j}-\langle I_0\rangle]\right]^{1/2}\left[\dfrac{1}{n}\displaystyle\sum_{j=1}^{n}[I_{1j}-\langle I_1\rangle]\right]^{1/2}} \tag{3.62}$$

where: I_{0j} is the initial intensity recorded by the j detector, and I_{1j} is the intensity recorded by the j detector when the sample surface is rotated an angle $\delta\theta$.

$$n = 1, 2, \ldots, 2\Delta_I \qquad (2\Delta_I < 512)$$

$$\langle I_0\rangle = \frac{1}{512}\sum_{h=1}^{512}I_{0h}$$

$$\langle I_1\rangle = \frac{1}{512}\sum_{h=1}^{512}I_{1h}$$

Figure 3.13 shows the speckle pattern intensity along a line of photodetectors. Changes in the same speckle pattern due to different illuminating angles can be seen.

The roughness of electroeroded brass samples was measured by the speckle correlation method, and then it was compared with the measurements made with a variable reluctance instrument. A good agreement was found between the theoretical model and experimental results.

Brass samples with 0.5 to 100% of electroeroded surface, that is, from 0.5% of the surface with pits of electroerosion (with a surface roughness very similar to that

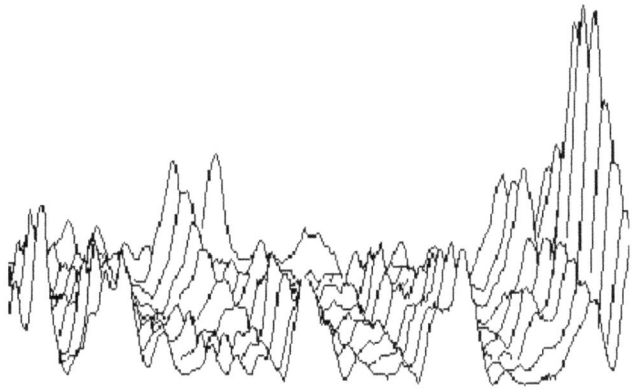

FIGURE 3.13 Speckle patterns displaced some pixels for each different illuminating angle.

of the noneroded surface, $\sigma \cong 1 \ \mu$m) up to 100% erosion ($\sigma \cong 7 \ \mu$m) were studied. An erosion process of only 5% is readily observable with this method. Figure 3.14 shows the experimental speckle intensity correlation ρ_I versus $\delta\theta$ evaluated for different stages of electroerosion.

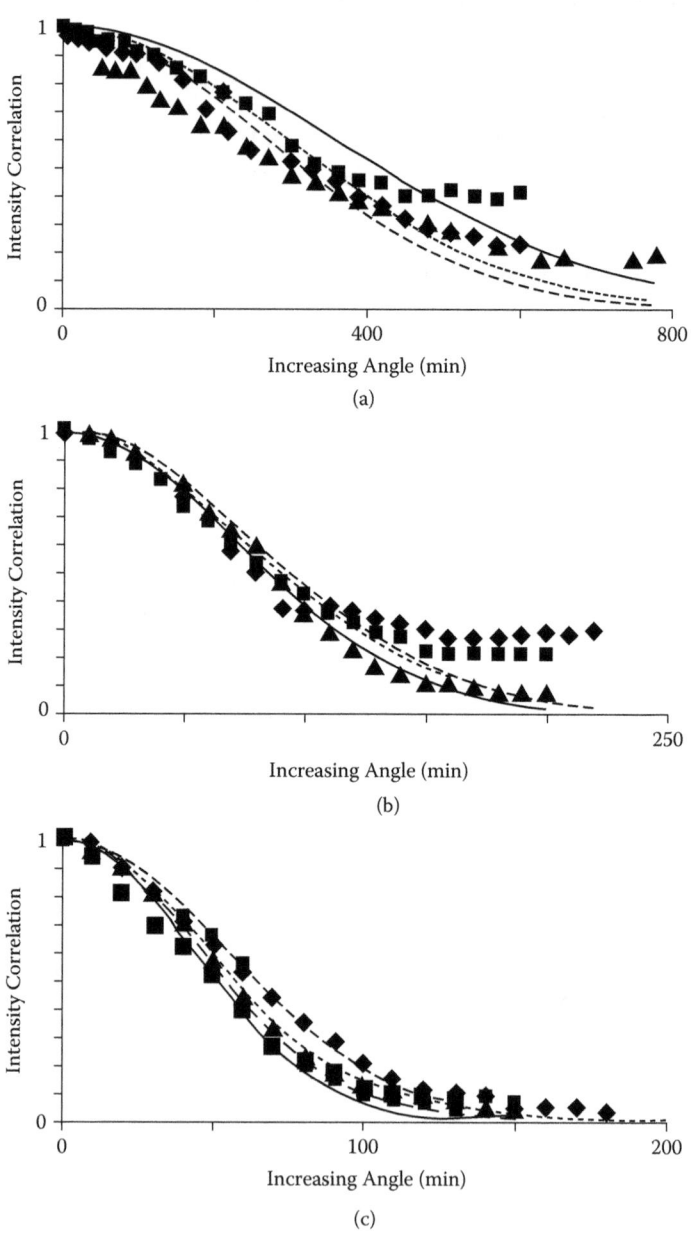

FIGURE 3.14 Correlation of speckle intensity C_1 versus the increasing angle $\delta\theta$ (a) $\square = 0\%$, o $= 1\%$, $\Delta = 5\%$; (b) $\square = 10\%$, o $= 20\%$, $\Delta = 40\%$; and (c) $\lozenge = 50\%$, $\square = 60\%$, $\Delta = 80\%$, o $= 90\%$, $\square = 100\%$.

Figure 3.14a shows that the curves of surfaces without or with only a few corrosion pits depart from the Gaussian theoretical curve. The rough surface is also different from the Gaussian height distribution case. When the corrosion is greater, the rough surface resembles more closely a Gaussian surface and there is a better fit between experimental and theoretical results (Figure 3.14b,c).

3.2.3.4 Analysis of Experimental Results

In order to determine the amount of the eroded area, we have taken pictures of the samples with a bidimensional CCD camera. In the first picture (Figure 3.15a) we have registered the surface illuminated by laser. The second one (Figure 3.15b) was obtained in the same conditions but without laser illumination (rectangle in Figure 3.15b). In this case, the well diameter is about 105 μm.

Errors in the calculation of s are principally due to the definition of the edge of the electroerosion well. In Figure 3.15b, it can be seen that the red rectangle introduces an additional error.

With the same program, the rectangular photograph was converted into two gray tones (see Figure 3.16), and then the number of dark and white pixels was counted, obtaining the percentage of attacked surface.

In Figure 3.17, the micrograph of one of the electroeroded wells can be seen. The little cross show the nonaffected zones, which are the reason for the lack of precision in the edge definition.

The developed method is only valid when there is no overlapping of wells. We have considered that up to 50% of attacked area has no superposition. So, we have represented the roughness square versus the ratio of the eroded area up to $s = 0.5$, as shown in Figure 3.16, obtained by linear regression.

From the analysis of the experimental results and the curve of Figure 3.18, we can conclude that the variation in the roughness in terms of s satisfies Equation 3.41. From the fitting parabola, we have deduced the following parameters:

$$p = (8.2 \pm 0.5)\ \mu m$$

$$\sigma_1 = (1.1 \pm 0.1)\ \mu m \qquad\qquad (3.63)$$

$$\sigma_2 = (4.6 \pm 2)\ \mu m$$

(a) 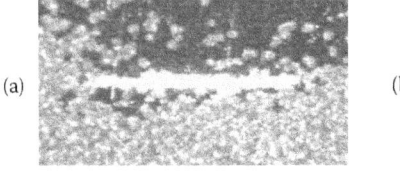 (b)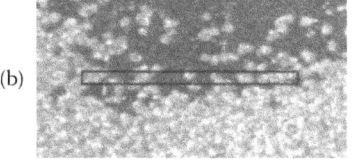

FIGURE 3.15 Image of (a) illuminated area, and (b) rectangle that approximates the illuminated area.

FIGURE 3.16 Levels of eroded surface with distinct gray tones: (a) and (b) 15% of eroded surface without any process and with only two gray tones, respectively; (c) and (d) 42%; (e) and (f) 60%.

The values obtained are very reasonable for both roughness and well depth considered in the model. In order to see the validity of this model, we have compared these results with those given by the variable reluctance instrument.

The samples of brass have zones of eroded surface of only 2 mm × 9 mm with an erosion degree from 0 to 100%. Due to the instrument evaluation length limitation, we cannot measure the roughness in terms of s. Nevertheless, it is possible to register a profile to measure the parameters p, and σ_2, determining the value of σ_1 in a nonattacked zone. The results obtained are

$$p = (8 \pm 2) \ \mu m$$

$$\sigma_1 = (0.5 \pm 0.2) \ \mu m \qquad (3.64)$$

$$\sigma_2 = (0.9 \pm 0.4) \ \mu m$$

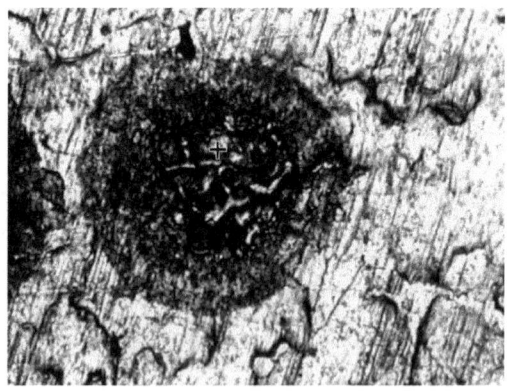

FIGURE 3.17 Photography of an electroeroded well with a magnification of 300X.

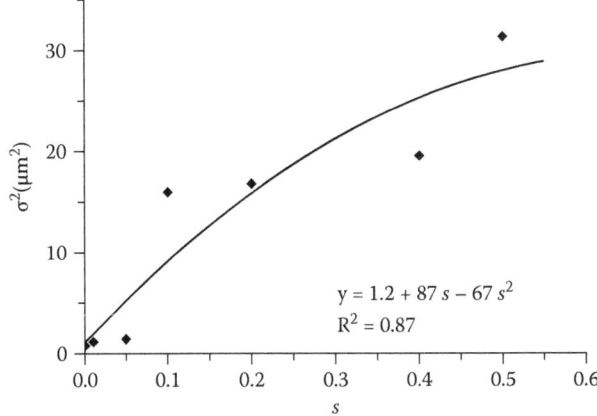

FIGURE 3.18 Roughness square versus ratio of electroeroded area where the diamonds are the experimental data, and the line is the fitting curve.

Even when the error in these parameters is about 50%, it is possible to compare them with those from the speckle correlation method.

As σ_2 depends on the other two parameters p and σ_1 (Equation 3.42), it is difficult to determine its value. In Equation 3.58, it can be seen that σ_2 is one order smaller than p, and it is about the order of the error in p. Therefore, the error in the determination of σ_2 is very important.

With the results reached up to now, a quadratic expression can be used to relate σ and s. Therefore, the proposed simple model can be a good tool to analyze the surface corrosion state.

3.2.4 APPARENT ROUGHNESS

Following Beckmann,[32] the mean scattered intensity from a translucent rough surface will be (Figure 3.19)

$$\langle I(\theta_2)\rangle = \langle E_2 E_2^*\rangle = \frac{|K_{2R}|^2 2LT\sqrt{\pi}}{v_z \sigma} \exp\left(-\frac{v_x^2 T^2}{4v_z^2 \sigma^2}\right) \qquad (3.65)$$

where, as usual, the vector $\bar{v}$ reports the optical path difference due to the surface. Therefore, for the transmission geometry (Figure 3.19), the Cartesian components of the vector are

$$v_x = k \ (n\,sen\theta_t - n_0 sen\theta_2) = k \ (n_0 sen\theta_1 - n_0 sen\theta_2) = k \ n_0 (sen\theta_1 - sen\theta_2) \qquad (3.66)$$

$$v_z = k \ (n \cos\theta_t - n_0 \cos\theta_2) = k \ (\sqrt{n^2 - n_0^2 sen^2\theta_1} - n_0 \cos\theta_2) \qquad (3.67)$$

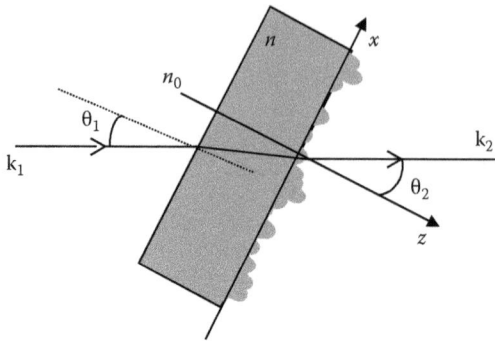

FIGURE 3.19 Transmission through one material.

Let us suppose that the same surface is immersed in a liquid having n_2 refractive index (Figure 3.20). In this situation, the mean scattered intensity will be

$$\langle I'(\theta_2) \rangle = \langle E_2' E_2'^* \rangle = \frac{|K_{2R}|^2}{v_z'} \frac{2LT\sqrt{\pi}}{\sigma} \exp\left(-\frac{v_x'^2 T^2}{4v_z'^2 \sigma^2} \right) \qquad (3.68)$$

Where the immersion liquid has introduced another phase difference. Vectors $\bar{v}' \neq \bar{v}$ will be different (Figure 3.20)

$$v_x' = k(n\,\text{sen}\,\theta_t - n_2\text{sen}\,\theta_t') = k\,(n_0\text{sen}\,\theta_1 - n_0\text{sen}\,\theta_2) = k\,n_0\,(\text{sen}\,\theta_1 - \text{sen}\,\theta_2) = v_x \qquad (3.69)$$

$$v_z' = k\,(n\,\cos\theta_t - n_2\,\cos\theta_t') = k\,(\sqrt{n^2 - n_0^2\text{sen}^2\theta_1} - \sqrt{n_2^2 - n_0^2\text{sen}^2\theta_2}) \qquad (3.70)$$

It can be seen that $v_x' = v_x$, but it is not true for the z component.

A variation in the refractive index produces a change in the optical path and therefore a phase modification. This phase variation can be interpreted as a change in the surface structure and may be due to changes in T and σ.[24] Thus, the mean scattered

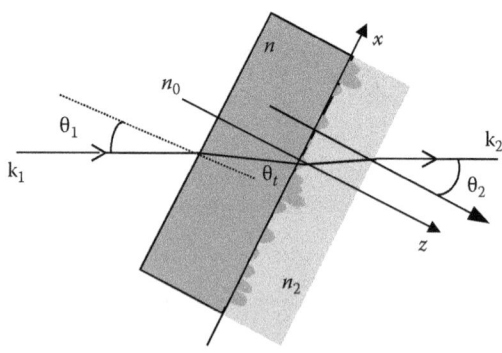

FIGURE 3.20 Transmission through two materials.

intensity by the surface immersed in a liquid is similar to that scattered by another surface with parameters T' and σ'.

For the sake of simplicity, $\theta_1 = \theta_2 = 0$, and $T' = T$; only the variation in the roughness σ is considered. The light is scattered as if the surface roughness were

$$\sigma' = \frac{v_z'(\theta_1 = \theta_2 = 0)}{v_z(\theta_1 = \theta_2 = 0)}\sigma \tag{3.71}$$

From Equations 3.66, 3.67, 3.69, 3.70, and 3.71, the apparent roughness σ can be obtained *as*

$$\sigma' = \frac{n - n_2}{n - 1}\sigma \tag{3.72}$$

For this expression, the presence of a diffuser, that is, $n \ne 1$, was supposed. If $n = 1$, the incident plane wave front continues on its way, and the expression (Equation 3.72) diverges, but this case has no physical meaning.

Finally, the mean intensity (Equation 3.68) is like the mean scattered intensity from a surface with roughness σ'. Then, the mean intensity is

$$\langle I'(\theta) \rangle = \langle E_2 E_2^* \rangle = \frac{|K_{2R}|^2 2LT\sqrt{\pi}}{v_z \sigma'} \exp\left(-\frac{v_x^2 T^2}{4 v_z^2 \sigma'^2}\right) \tag{3.73}$$

3.2.5 Theoretical Model for Blood Flow

In 1980, Bonner and Nossal[34] published a specific theoretical model for quasielastic scattering in blood cells, including the effect of surrounding stationary or very slowly moving blood vessel walls. It was intended to be used for small vessels.

The theory included the possibility of taking into account scattering in several moving cells, and diffuse scattering in the walls of the vessels. They found that the first moment of the spectrum of the scattered light was proportional to the root mean square (rms) speed of the cells V.

$$\langle \omega \rangle = \frac{\langle V^2 \rangle^{1/2} \beta}{(12\xi)^{1/2} a} f(\bar{m}) \tag{3.74}$$

with a the radius of the (assumed spherical) moving cell, ξ an empirical factor related to the actual shape of the cells, and β an instrumental factor. The variable $f(\bar{m})$ is a function of the average number of collisions $\bar{m}$ of a photon with moving cells.

It is linear with blood flow volume when $\bar{m} \ll 1$, and varies with the square root of blood flow volume when $\bar{m} \gg 1$.

Bonner and Nossal assumed that moving red blood cells could be approximated by spherical particles illuminated by a diffuse source. The validation presented the ability of the model to predict the blood volume flow; nevertheless, the amplitudes of features such as flow speed and collision number were far from the actual values.

3.3   NUMERICAL MODELS

In this section, we briefly describe some numerical models that have been used to simulate the dynamic speckle phenomenon as an alternative way to the theoretical models that are mainly concentrated on nonbiological simulations.

In the absence of accurate standards of activity, numerical simulations have been used to test for the behavior of different measuring algorithms. Some of them, which are intended mostly for general purposes, are going to be reviewed in this chapter.

They are based on approximations of the scattering phenomenon by discrete sets of scattering centers, followed by light propagation in the assumed experimental conditions in scalar diffraction approximation.

We will start with the simplest one[35] and then proceed to others, a little more elaborated and recently reported. Both free propagation and imaging geometries will be included. Multiple scattering and turbulence models will not be treated here.

Several numerical models have been developed following different approaches intending to reproduce the speckle patterns in time under simplifying assumptions, as they can be originated by the effect of multiple phenomena occurring in the surface, near the surface, in the skin of biological materials, or in the bulk of some industrial materials. These phenomena cause the Doppler effect from particles in movement, local variations in the refractive index, polarization changes, and diffuse scattering produced by complex organizations, particularly in biological tissue.

Some of these models[36,37] are based on free propagation using the basic concepts of the formation of the speckle pattern, the phase being the main origin of the changes of patterns in time. Two different ways to introduce the phase changes have been presented to simulate the formation of the dynamic interference pattern and its changes. The comparison with actual experimental data showed that the models reproduce several features of the experiments and previous theoretical knowledge of the simplified dynamics.

The approaches will be treated here as two models for the movement of the scatterers and one model of phase variation. The other two accounts presented in the literature[38,39] used Fraunhoffer subjective speckle. Those models will be discussed in this section with their main concepts, validation procedures, and results.

3.3.1   RANDOM-WALK SCATTERERS MODEL

In the random-walking scatterers model (RWS),[35] one assumes that the speckle is produced by a large number of rectangular slit-like scatterers performing a random walk on the surface of the sample. From this assumption, a dynamical diffraction pattern is calculated, assuming the Fraunhoffer diffraction. The step size of the random walk is related to the activity of the speckle pattern. The intensity value of a point in the observation plane can be expressed as

$$I(u,t) = \left| \frac{1}{2\pi} \int_{-\infty}^{\infty} R(x,t) \exp(-ikux) dx \right|^2 \tag{3.75}$$

where t is time, x the position in the one-dimensional diffuser, k the wave number, u the coordinate in Fourier space, and R represents the reflectance function of the scatter center, represented here by a slit. R can be expressed in this model as

$$R(x,t) = \sum_{i=1}^{N_s} \Pi_{b_i}[x - a_i(t)] \tag{3.76}$$

where

$$\Pi_b(x) = \begin{array}{ll} 1, & \text{if } -b < x < b \\ 0, & \text{elsewhere} \end{array} \tag{3.77}$$

$$a_i(t + \Delta t) = a_i(t) + S_i \tag{3.78}$$

where N_f is the number of slits, b_i its width, and a_i its location. S_i is the step size, restricted to an interval and limited by the presence of the other slits. Step size is an adjustable parameter of the model and corresponds to a global measure of the biological activity of the tissue. A tissue that is less active has organelles moving with lower speed and conversely. The model was adjusted using data obtained by electronic microscopy of a tissue (a seed in this case), and the model was validated by comparison with actual seed's time history speckle pattern (THSP) images under different moisture degrees.

It can be seen that, in spite of the simplicity of the model, it captures the dynamics as to show a good fit to the correlation functions of the seeds with different states of humidity, which is seen in Figure 3.21.

3.3.2 SCATTERERS' MOVEMENT MODEL[36]

This model adopts a simulation based on the composition of the THSP matrix, building each column considering the contributions of all the points in the illuminated

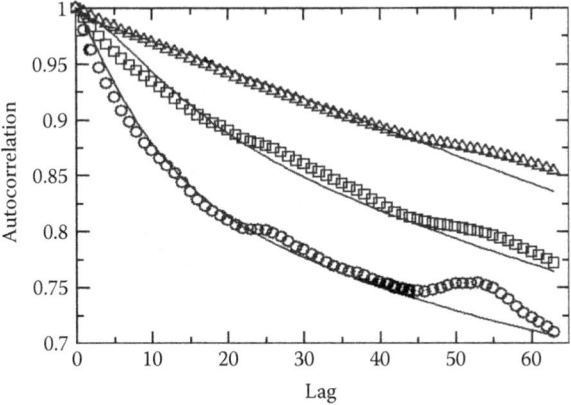

FIGURE 3.21 Autocorrelation curves of a seed with three moisture levels and the simulated results in the continuous line. (From Nascimento, A.L. et al., *Rev. Ciên. Agrotec.*, 31, 456, 2007. With permission.)

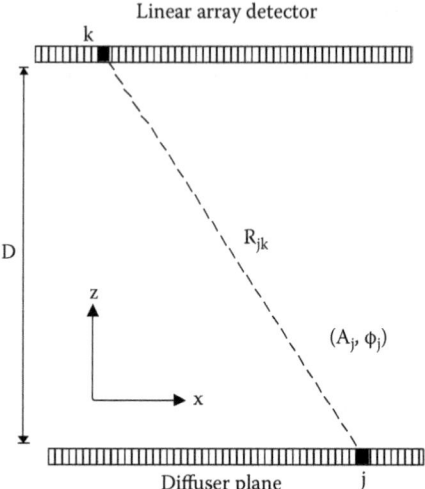

FIGURE 3.22 Schematic setup used in the simulation: free propagation between a diffuser plane and the detector plane. (Reprinted from *J. Opt. A: Pure Appl. Opt*, Numerical model for dynamic speckle: An approach using the movement of the scatterers, Héctor J Rabal, Ricardo Arizaga, Nelly L Cap, Eduardo Grumel and Marcelo Trivi, 2003. With permission of IOP Publishing, Edinburgh, U.K.)

plane in Fresnel free propagation. Both two-dimensional and one-dimensional diffusers were simulated, with the one-dimensional diffuser version reduced to a line with the same size of the detector column. The light field was calculated by adding all the contributions as spherical waves using the expression

$$U(1) = \sum \frac{A}{R} \exp(ikR + \phi) \tag{3.79}$$

where $k = \frac{2\pi}{\lambda}$, and R is the distance between the scattering center j and the pixel l, in accordance with Figure 3.22.

Considering the D value much greater than the size of the detector as well as the size of the diffuser, it is possible to assume that $R \cong D$ in the denominator with negligible error.

The amplitude A and its phase are randomly and uniformly distributed in [0, 1] and $[-\pi, \pi]$ intervals, respectively.

Once the first column is formed, the next step is the construction of the subsequent column considering the time evolution of the speckle. This evolution was considered in accordance with Equation 3.2, where $v(j)$ is the velocity assigned to the scattering center j, R_0 being the initial position.

$$R(j,k,t) = R_0(j,k) + v(j) t \tag{3.80}$$

Uncorrelated movement of the scattering centers in the z direction was simulated by setting the velocities to the values

$$v(j) = \varepsilon \, r(j) \tag{3.81}$$

ε being the velocity amplitude, and $r(j)$ a random variable with chosen statistics in the [0, 1] interval.

In the simulations, the value of ε determines the maximum velocity of the scattering centers (in millimeters per time interval between two consecutive frames).

The spatial simulations obtained with such a simple model agree quite well with the first-order statistics of experimental speckle patterns, both for the amplitude and the phase. For the simulation of dynamic speckle patterns, the temporal histories thus generated are in reasonable resemblance with experimental ones.

Figure 3.23 shows the result of the simulation with respect to the scatterers' velocity, by using the inertia moment (IM) measurement.

By using this simple model, several properties are found to qualitatively and quantitatively agree with experimental results. The moment of inertia increases monotonically with the ε value (Figure 3.23), and the spatial contrast shows the theoretically predicted value.

A two-dimensional image was constructed with the model procedure considering different proportions of moving and still-scattering centers.

$$\frac{\langle\sigma^2\rangle_x}{\langle I\rangle^2} = 1 - (1-\rho)^2 \tag{3.82}$$

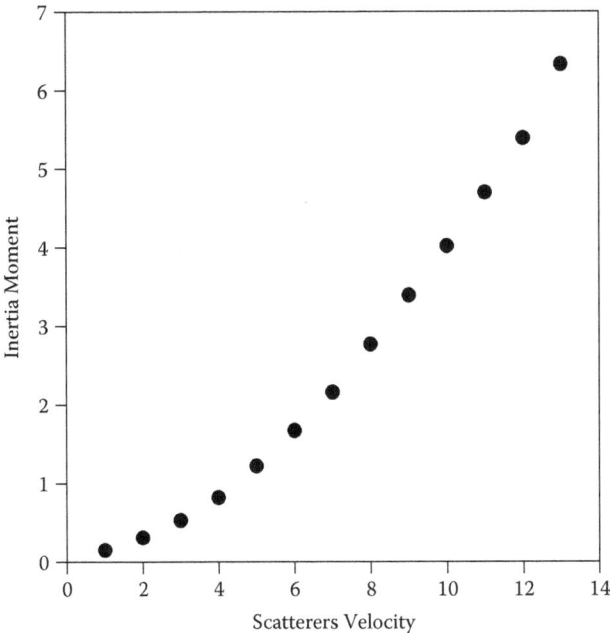

FIGURE 3.23 Result of the simulation: Inertia moment measurement as a function of scatterers velocity. (Reprinted from *J. Opt. A: Pure Appl. Opt*, Numerical model for dynamic speckle: An approach using the movement of the scatterers, Héctor J. Rabal, Ricardo Arizaga, Nelly L. Cap, Eduardo Grumel and Marcelo Trivi, 2003. With permission of IOP Publishing, Edinburgh, U.K.)

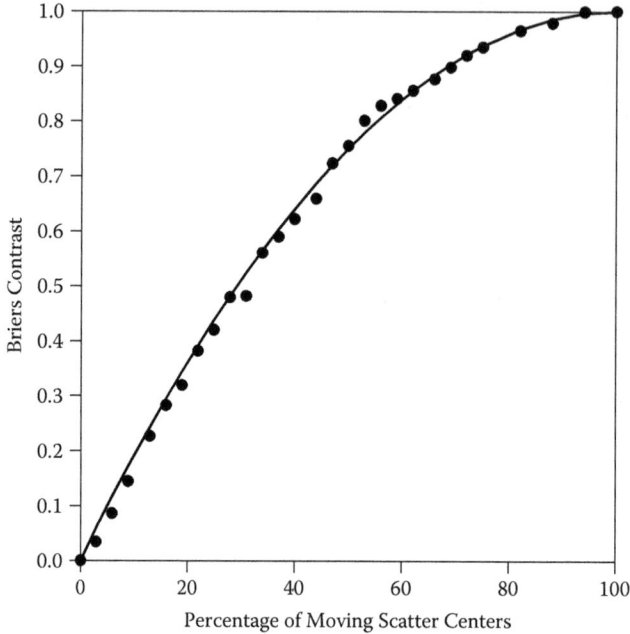

FIGURE 3.24 Briers (spatial) contrast as a function of the percentage of moving scattering centers. Solid line shows the theoretical prediction and the points the numerically simulated result. (Reprinted from *J. Opt. A: Pure Appl. Opt*, Numerical model for dynamic speckle: An approach using the movement of the scatterers, Héctor J. Rabal, Ricardo Arizaga, Nelly L. Cap, Eduardo Grumel and Marcelo Trivi, 2003. With permission of IOP Publishing, Edinburgh, U.K.)

where we call the left side of the equation Briers contrast; and σ is the time standard deviation of the intensity, with the brackets indicating spatial mean value; and ρ being the ratio of the mean intensity of light from the moving scattering centers to the mean total intensity of scattered light I.[40]

The results are shown in Figure 3.24 along with the theoretical result of Briers contrast (in solid line). It can be seen that the obtained speckle contrast as a function of the ratio of moving to stationary scattering centers is in good agreement with the formula given by Briers for fully developed speckle patterns.

3.3.3 PHASE VARIANCE MODEL

This model was based on the image of subjective dynamic speckle patterns being a two-dimensional diffuser producing a set of speckle patterns that are simulated in the image plane of a lens, and then a THSP image following the evolution.

The intensity of light reaching the CCD was simulated by using the expression[41–43]

$$I(m,n,k) = \left| FFT^{-1}[H[FFT(\exp(i\phi(m,n,k)])] \right|^2 \qquad (3.83)$$

where

$m = n = 0,1,2, \ldots ,N - 1$ the pixel coordinates of the CCD, and

$\phi(m,n,k)$ the phase distribution matrix of the scattered light from the surface.

Fast Fourier Transform (FFT) and its inverse were used to implement the low-pass filter H with radius r in the Fourier space representing the aperture of the camera. The next step is to create a sequence of speckle matrices with changes in the pattern produced by the change of the phase represented by $\phi(m,n,k)$. Equation 3.84 represents the phase $\phi(m,n,k)$ varying in accordance with a correlation coefficient function built considering a random movement of the scatterers in two directions (Figure 3.25). The phase can be described as

$$\phi(m,n,k) = \phi(m,n,k-1) + G(m,n,k)\sigma[\Delta\phi(m,n,k-1,k)] \qquad (3.84)$$

where σ represents the standard deviation of the phase between $k - 1$ and k, with k varying from 1 to $K - 1$. In this equation, G is an $N \times N$ random matrix with Gaussian distribution (with zero mean, and standard deviation equal to 1). After mathematical manipulations, the phase is represented as

$$\phi(m,n,k) = \phi(m,n,k-1) + G(m,n,k)\sqrt{\ln c(k-1) - \ln c(k)} \qquad (3.85)$$

c being the correlation coefficient function expressed in Equation 3.86:

$$c(k) = \frac{\langle I(0)I(k)\rangle - \langle I(0)\rangle\langle I(k)\rangle}{[(\langle I^2(0)\rangle - \langle I(0)\rangle^2)(\langle I^2(k)\rangle - \langle I(k)\rangle^2)]^{1/2}} \qquad (3.86)$$

Several results obtained with the model were found to agree with the experiments.

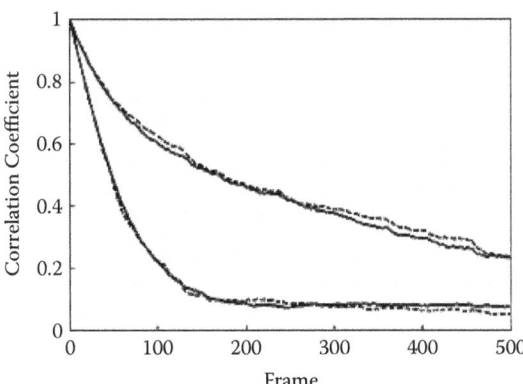

FIGURE 3.25 Comparison between the temporal variations of the correlation coefficient determined from the simulated speckle patterns (dashed line) and the process of paint drying (continuous line) for two different paints. (Reprinted from *Opt. Commun.*, 260, A. Federico, G. H. Kaufmann, G. E. Galizzi, H. Rabal, M. Trivi, and R. Arizaga, Simulation of dynamic speckle sequences and its applications to the analysis of transient processes, 493–499. With permission of Elsevier Science.)

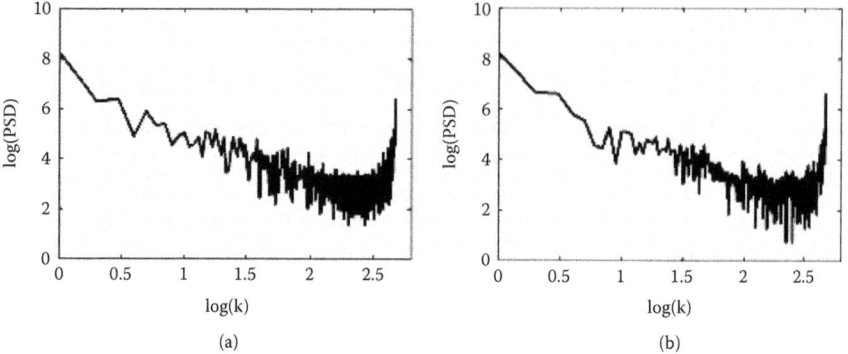

FIGURE 3.26 (a) Power spectral density of the temporal evolution of the intensity at an arbitrary pixel obtained from a process of paint drying, and (b) sequence of simulated speckle patterns. (Reprinted from *Opt. Commun.*, 260, A. Federico, G. H. Kaufmann, G. E. Galizzi, H. Rabal, M. Trivi, and R. Arizaga, Simulation of dynamic speckle sequences and its applications to the analysis of transient processes, 493–499. With permission of Elsevier Science.)

Due to the particular way in which phase is calculated, including only the movement of the scattering centers, this model can be expected to perform well in phenomena characterized by growing or eroding material in a surface, such as corrosion and efflorescence.

Another contribution[39] proposed improvements in this model in order to consider more general situations, especially those of being free from the focusing condition.

If a single lens image formation scheme is used (Figure 3.27), the intensity pattern registered at the image plane, including the possibility of defocused situations, is related to the optical field at the object plane by the following expression in the scalar diffraction approximation:

$$\left| U(u,v) \right|^2 \propto \left| F\left\{ P(\lambda d_i \tilde{x}, \lambda d_i \tilde{y}) e^{j\frac{k}{2}\left(\frac{1}{d_i}+\frac{1}{d_o}-\frac{1}{f}\right)(\lambda d_i)^2(\tilde{x}^2+\tilde{y}^2)} F\left\{ U\left(\frac{d_o}{d_i}\tilde{\xi}, \frac{d_o}{d_i}\tilde{\eta}\right) e^{j\frac{k d_o}{2 d_i^2}(\xi^2+\bar{\eta}^2)} \right\} \right\} \right|^2$$

(3.87)

where $U(\ )$ is the optical field, $F\{\ \}$ is the Fourier transform, $P(x,y)$ is the lens pupil function, f is the lens focal distance, and d_o and d_i are the distances from the lens to the object and image plane, respectively.

For simulating speckle patterns in a computer, the previous expression must be written in a discrete version. By using the discrete Fourier transform (DFT), it becomes

$$\left| U[g,h] \right|^2 \propto \left| \mathrm{DFT}\left\{ P[k,l] e^{j\pi\lambda d_o^2\left(\frac{1}{d_i}+\frac{1}{d_o}-\frac{1}{f}\right)\left[\left(\frac{k}{M_\xi \Delta_\xi}\right)^2+\left(\frac{l}{M_\eta \Delta_\eta}\right)^2\right]} \mathrm{DFT}\left\{ U[m,n] e^{j\frac{\pi}{\lambda d_o}[(m\Delta_\xi)^2+(n\Delta_\eta)^2]} \right\} \right\} \right|^2$$

(3.88)

where $P[k,l] = P(k\Delta x, l\Delta y)$; $U[m,n] = U(m\Delta x, n\Delta h)$; $U[g,h] = U(g\Delta u, h\Delta v)$; and $M\xi$ and $M\eta$ are the image coordinates considering magnification M.

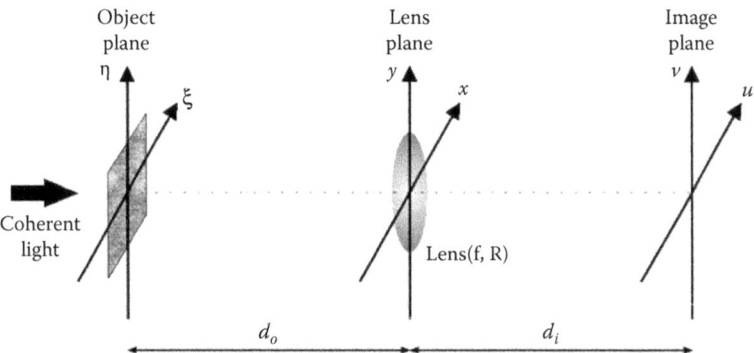

FIGURE 3.27 Simulated optical setup.

In order to validate this model, two theoretical models were used considering translational and boiling speckle simulations, and simulations of static and moving scatterers.

3.3.3.1 Translational and Boiling Speckle Simulations

When the object plane is in movement, the observed speckle becomes a dynamic pattern that may or may not replicate the movement of the object. If each speckle grain reproduces the object movement, the speckle is called *translational speckle*.

On the other hand, if no object movement can be inferred from the dynamic speckle pattern, then the speckle pattern is called *boiling speckle*.

Okamoto and Asakura[44] analyzed the space–time autocorrelation function of the intensity and found that the conditions for pure translational and pure boiling speckle are

pure boiling speckle: $d_i = f$ (3.89)

pure translational speckle: $\rho_0 = -d_0$ (3.90)

where ρ_0 is the radius of curvature of the incident light wave front.

3.3.3.2 Simulations of Static and Moving Scatterers

These simulations assume that the object is composed of both static and uniformly moving scatterers. The uniform movement is represented as constant variations in the phase of each scatterer along the image sequence. The initial phase of each scatterer and the corresponding phase variation, if it is in movement, are chosen as random numbers. Briers[40,45] found the following relationships between contrast (temporal or spatial) and the percentage of moving scatterers (Figures 3.28 and 3.29).

$$\rho = 1 - \left[1 - \frac{\langle \sigma_t^2(x,y)\rangle}{\langle I\rangle^2} \right]^{1/2} \qquad \rho = 1 - \frac{\sigma_I(x,y)}{\langle I\rangle} \qquad (3.91)$$

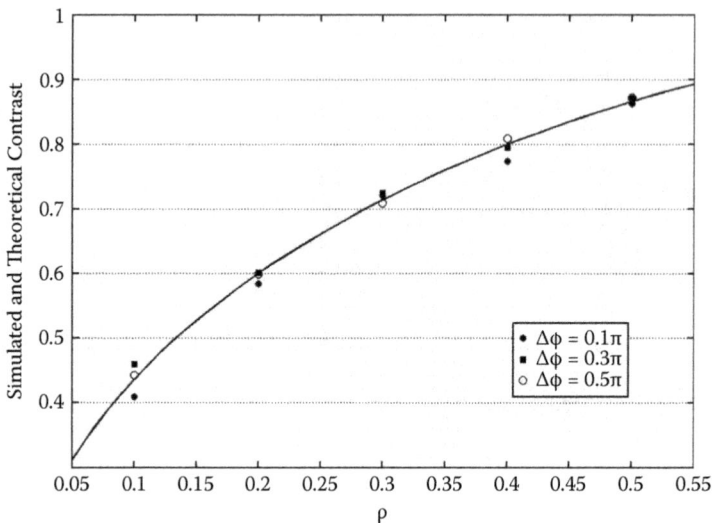

FIGURE 3.28 Simulated and theoretical (spatial) contrast of the time-integrated dynamic speckle pattern for different percentages of moving scatterers.

This model shows dynamic speckle simulations as an interesting tool to understand scattering phenomena produced when coherent light illuminates a diffuse object.

Simulation performances were successfully validated by the reproduction of Briers and Okamoto–Asakura statistics of dynamic speckles The model was also tested by

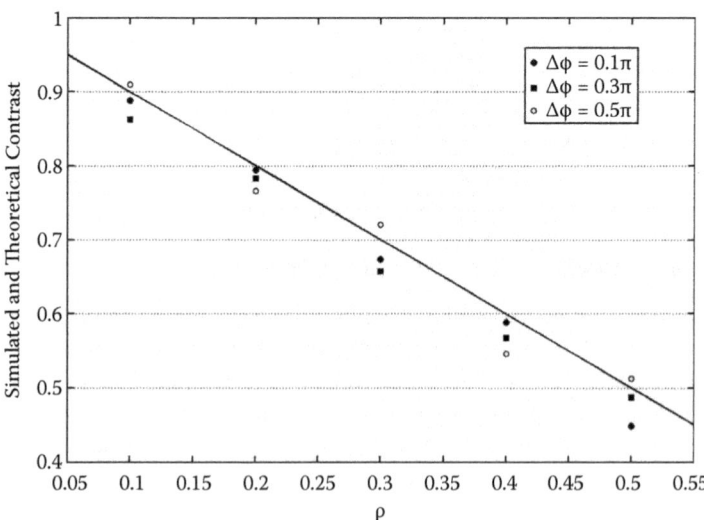

FIGURE 3.29 Simulated and theoretical (temporal) contrast for different percentages of moving scatterers.

comparing the real and simulated power spectrum density of patterns generated by vitreous and floury maize endosperm, and its results showed similar values.

3.3.4 COPULA MODEL

Recently, a model that uses the statistical concept of copula[37] to generate time-evolving speckle patterns with prescribed correlations was proposed. In its present version, it can be used to generate objective speckles.

It starts with the generation of two independent uniformly distributed random variables (RVs). Through the Box–Mueller transformation, they are transformed into two standard normally distributed RVs.

A scaling and rotation operation on those variables gives bivariate RVs with prescribed correlation coefficient r. Finally, they use the percentile transformation to obtain a pair of RVs uniformly distributed and with prescribed correlation.

The RVs obtained in this last step are then used as the phases ϕ_1 and ϕ_2 in a simulation of a pair of free propagation speckle patterns by using the FFT.

The actual correlation obtained between these two speckle patterns is

$$\rho = \exp\left(-\sigma_{\Delta\phi}^2\right) \tag{3.92}$$

where

$$\sigma_{\Delta\phi}^2 = \mathrm{var}(\phi_1 - \phi_2) \tag{3.93}$$

is the variance of the phase difference, as results from the complex Gaussian moment theorem. From the desired correlation ρ, the required values of r are traced back.

This algorithm gives continuous-phase trajectories that also produce continuous and realistic time evolutions of simulated speckle patterns. Exponentially correlated and Gaussian correlated sets of objective speckle frames are simulated in this way.

3.4 CONCLUSIONS

Some models were presented in this chapter showing their abilities to reproduce the complex phenomena related to the interaction of light, and biological and nonbiological matter. The theoretical model showed the reliability to reproduce the changes in an electroeroded surface using the dynamic speckle point of view. This development unfolded the opportunity to do the same to other nonbiological material.

The development of numerical models are an interesting alternative to study the behavior of the dynamic speckle in more complex structures such as biological matter.

By means of the developed theoretical model, a decreasing quadratic exponential expression for the speckle correlation coefficient in the function of roughness variation has been found. Experimental results show that this coefficient can be used to analyze samples undergoing a surface process.

Some recent numerical models were proposed to describe free propagation and image plane dynamic speckle patterns. These models, in spite of their simplicity, reproduce the statistics well, and the dynamics and previous theoretical results reasonably well.

The first one, based in a random walk model, shows a good fit with the correlation functions of seeds with different states of moisture.

The next model should be valid within the assumptions that the detector is supposed to be within the Fresnel region of propagation of the light coming from the scattering centers. The latter are assumed to be spherical so that they contribute in equal amounts to each detector and to be moving in only one direction. The refractive index is assumed to be constant in time. It reproduces well the first- and second-order statistics and Briers theoretical result for the proportions of moving scatterers.

Another simple numerical model to simulate a temporal sequence of two-dimensional dynamic speckle patterns was described on the basis of the space variance of the phase. It considers only the movement of the scattering centers and produces dynamic speckle patterns that verify the correlation coefficient obtained from the experiments, provided that it is decreasing, monotonously and the power spectral density of the scattering intensity.

The observed results reflect the spatial variance of the phase determined from the spatial autocorrelation of the intensity of consecutive frames in a sequence of dynamic speckle patterns. In some simplified nonbiological examples such as corrosion, the variance of phase can be interpreted as due to temporal changes in the height of the surface, a well-defined physical property, but biological processes are not so simple, and some ambiguity could be expected. The measured magnitude can be associated with the velocity of the scattering centers or to refractive index variations or to other physical features of the surface that could be known a priori.

The use of the concept of copula to generate two representative functional forms for the decorrelation of an objective, Gaussian, and exponential speckle sequence was also commented upon. This approach is not limited to the monotonically decreasing functions such as the latter. More complex dependencies could be simulated, including the more frequent Voigt profile that is obtained when both Lorentzian and Gaussian broadenings are simultaneously present.

The field is still evolving from very simple geometrical situations and discrete phase steps to more realistic experimental setups and continuous-phase variations.

It is to be expected that new approaches will continue to be reported that could be of help in finding some type of standards and giving deeper insight into the origins of the dynamics of speckle patterns.

REFERENCES

1. Bennett, J. M., Recent developments in surface roughness characterization, *Meas. Sci. Technol.*, 3, 1119, 1992.
2. Rao, C. B. and Raj, B., Study of engineering surfaces using laser-scattering techniques, *Sadhana—Proceedings in Engineering Sciences*, Springer India, 28, 739, 2003.
3. Hagemaier, D. J., Wendelbo, A. H., and Bar-Cohen, Y., Aircraft corrosion and detection methods, *Mat. Eval.*, 43, 426, 1985.

4. Russell, S. S., Sutton, M. A., and Chen, H. S., Image correlation quantitative non-destructive evaluation of impact and fabrication damage in a glass fiber-reinforced composite system, *Mat. Eval.*, 47(5), 550, 1989.

5. Newman, J. W., Shearographic inspection of aircraft structure, *Mat. Eval.*, 49(9), 1106, 1991.

6. Kaufmann, G. H. and Baumann, C. E., Residual-deformation evaluation using speckle photography, in *Proc. of the Pan American Congress of Applied Mechanics*, Rio de Janeiro, 1989, 213.

7. Buerkle, J. et al., Rapid defect detection by laser light scattering, *Mat. Eval.*, 50(6), 670, 1992.

8. Hageniers, O. L., Diffracto sight—a new form of surface analysis, *Proc. SPIE*, 814, 193, 1987.

9. Reynolds, R. L. and Hageniers, O. L., Optical enhancement of surface contour variations, *Proc. SPIE*, 954, 208, 1988.

10. Komorowski, J. P., Simpson, D. L., and Gould, R. W., Enhanced visual technique for rapid inspection of aircraft structures, *Mat. Eval.*, 49(12), 1486, 1991.

11. Gaggioli, N. G. and Roblin, M. L., Étude des états de surfaces par les propietés de diffusion à l'infini en lumière transmise, *Opt. Comm.* 32, 209, 1980.

12. Sthel, M. S. et al., Speckle patterns direct photographic correlation for measuring surface roughness, *Proc. SPIE*, 813, 561, 1987.

13. Rebollo, M. A. et al., Determinaciones no destructivas por el método de correlación de speckles, in *Actas I RIAO*, Barcelona, 1992, 382.

14. Muramatsu, M., Guedes, G. H., and Gaggioli, N. G., Speckle correlation used to study the oxidation process in real time, *Opt. Laser Tech.*, 26, 167, 1994.

15. Quintián, F. P. et al., Relationship between speckle correlation and refraction index variations: Applications for roughness measurements, *Opt. Eng.*, 35(4), 1175, 1996.

16. Gale, M. F. R. et al., Digital speckle correlation for non-destructive testing of corrosion, in *Proc. 7th ECNDT*, Copenhagen, 1998, 717.

17. Hogert, E. et al., Estudio de la corrosion mediante el analisis de la intensidad media dispersada, in *III RIAO y VI OPTILAS*, Cartagena, 1998, AOIB 16.

18. Léger, D., Mathieu, E., and Perrin, J., Optical surface roughness determination using speckle correlation technique, *Appl. Opt.*, 14, 872, 1975.

19. Ruffing, B. and Anschutz, J., Surface roughness measurement by 2-D digital correlation of speckle images, *Proc. SPIE*, 814, 105, 1987.

20. Groh, G., Engineering uses of laser-produced speckled patterns, in *Proc. of the Symposium on the Engineering Uses of Holography*, Cambridge University Press, London, 483, 1970, 483.

21. Weigelt, G. P., Real time measurement of the motion of a rough object by correlation of speckle patterns, *Opt. Comm.*, 19, 223, 1976.

22. Paiva, R. D. et al., Study of the electroerosion process by the analysis of speckle correlation, *Insight*, 43, 235, 2001.

23. Yoshimura, T., Statistical properties of dynamic speckles, *JOSA*, 3, 1032, 1986.

24. Fang, Q. et al., A fast deformation analysis method by digital correlation technique, *Proc. SPIE*, 954, 333, 1988.

25. Rabal, H. J. et al., Transient phenomena analysis using dynamic speckle patterns, *Opt. Eng.*, 35, 57, 1996.

26. Ruth, B., Superposition of two dynamic speckle patterns: An application to non-contact blood flow measurements, *J. Mod. Opt.*, 34, 257, 1987.

27. Aizu, Y. and Asakura, T., Bio-speckle phenomena and their applications to the evaluation of blood flow, *Opt. Laser Technol.*, 23, 205, 1991.

28. Oulamara, A., Tribillon, G., and Duvernoy, J., Biological activity measurement on botanical specimen surfaces using a temporal decorrelation effect of laser speckle, *J. Mod. Opt.*, 36, 165, 1989.

29. Hinsch, K., Coherent optical metrology for environmental diagnostics, *Proc. SPIE*, 1524, 292, 1992.

30. Tanaka, S., Takenaka, I., and Ohtsuka, Y., Statistical evaluation of phase fluctuations of light scattered from apple peel, in *Optical Methods in Biomedical and Environmental Sciences*, Ohtzu, H. and Komatsu, S., Eds., Elsevier, Amsterdam, 1994, 15.

31. Papoulis, A., *Probability, Random Variables, and Stochastic Processes*, McGraw-Hill Kogakusha, Tokyo, 1965.

32. Beckmann, P. and Spizzichino, A., *The Scattering of Electromagnetic Waves from Rough Surfaces*, Pergamon Press, New York, 1963.

33. Goodman, J. W., *Introduction to Fourier Optics*, McGraw-Hill, San Francisco, 1968.

34. Bonner, R. and R. Nossal, Model for laser Doppler measurements of blood flow in tissue *Appl. Opt.*, 20(12), 2097, 1980.

35. Nascimento, A. L. et al., Desenvolvimento de um modelo para o biospeckle na análise de sementes de feijão (*Phaseolus vulgaris* L.), *R. Ciên. Agrotec.*, 31(2), 456, 2007.

36. Rabal, H. J. et al., Numerical model for dynamic speckle: An approach using the movement of the scatterers, *J. Opt. A: Pure Appl. Opt.*, 5, S381, 2003.

37. Duncan, D. D. and Kirkpatrick, S. J., The copula: A tool for simulating speckle dynamics, 25(1), *J. Opt. Soc. Am. A*, 231, 2008.

38. Federico, A. et al., Simulation of dynamic speckle sequences and its application to the analysis of transient processes, *Opt. Comm.*, 260, 493, 2006.

39. Sendra, G. H. et al., Simulations of dynamic speckle image activity, in *6th Iberoamerican Meeting on Optics and 9th Latin American Meeting on Optics, Lasers and Applications*, Campinas, Brazil, 2007.

40. Briers, J. D., The statistics of fluctuating speckle patterns produced by a mixture of moving and stationary scatterers, *Opt. Quant. Electr.*, 10, 364, 1978.

41. Davila, A., Kaufmann, G. H., and Kerr, D., Digital processing of electronic speckle pattern interferometry addition fringes, *Appl. Opt.*, 33, 1994, 5964.

42. Equis, S. and Jacquot, P., Simulation of speckle complex amplitude: Advocating the linear model, in *Proc. SPIE Speckle 06*, Bellingham, 2006, 6341.

43. Goodman, J. W., *Speckle Phenomena in Optics: Theory and Applications*, Roberts & Company, Englewood, Colo., 2007.

44. Okamoto, T. and Asakura, T., in *Progress in Optics XXXIV*, Wolf, E., Ed., Elsevier Science, Amsterdam, 1995, 183.

45. Briers, J. D., A note on the statistics of laser speckle patterns added to coherent and incoherent uniform background fields, and a possible application for the case of incoherent addition, *Opt. Quant. Electr.*, 7, 422, 1975.

4 # Methods of Dynamic Speckle Analysis
Statistical Analysis

Ricardo Arizaga

CONTENTS

4.1   INTRODUCTION

Dynamic speckle patterns analysis from active materials gives us much useful information, as introduced in Chapter 2. The challenge is then to correlate that information with the phenomenon to be monitored.

Links between the phenomenon and active objects under laser illumination started to be made in the 1970s as many new approaches were found due to the development of computers and optical devices such as charge coupled devices. The metrological uses of dynamic speckle were then consolidated by systematic testing of all the approaches suggested by the many applications. Measuring a biological or physical variable means to obtain from an illuminated object useful information based on a transduction, which allows the observer to evaluate, by processed images or numbers, the variables under control.

This chapter describes numerical approaches for analyzing the dynamic speckle patterns produced by biological and nonbiological materials under laser illumination. It is possible to start analyzing a technology presented in the 1970s using one speckle pattern,[1] introducing the spatial contrast index as a first step to further indices or methodologies that adopt the time domain in their calculus. Speckle contrast with respect to time, as a first order statistical, is followed by second order statistical methods

95

which are evaluated using time history speckle pattern (THSP) images, treated in detail in this chapter. Autocorrelation functions, statistical cummulants, and inertial moment methods are also described, with some actual examples for consideration.

4.2 SPATIAL CONTRAST

The assumptions concerning Gaussian statistics of speckle patterns lead to a decreasing exponential probability density function of intensity. One of the properties of such a distribution is that the standard deviation σ of the intensity is equal to the mean intensity:

$$\sigma = \langle I \rangle \tag{4.1}$$

In practice, speckle patterns often have a standard deviation that is less than the mean intensity, and this is observed as a reduction in the contrast of the speckle pattern. In fact, it is usual to define the speckle contrast as the ratio of the standard deviation to the mean intensity:

$$C = \sigma / \langle I \rangle \tag{4.2}$$

For the case of a single pattern of polarized speckle, the contract is close to one, and is called fully developed speckle pattern. Otherwise, the speckle pattern is called *partially developed*.

There are many causes for a decrease in speckle contrast. Reductions in the coherence of the light source or in the roughness of the surface are two. Another might be the addition to the speckle pattern of a uniform background of light, which may be either coherent with the speckle pattern or completely incoherent. Therefore, if the activity is low, the correlation between successive frames is high, and acquiring and accumulating frames is equivalent to adding incoherent but similar speckle patterns, so that the contrast stays high. Conversely, if activity is high, successive frames are very different and contrast falls abruptly. So, the contrast of the speckle pattern is roughly a measurement of the activity of dynamic speckle.

4.3 TEMPORAL CONTRAST

If the ratio $\sigma / \langle I \rangle$ is calculated from temporal signals recorded at a single detecting point on the speckle pattern, it does not reflect information of the velocity of moving speckle or the frequency components of intensity fluctuations, but it can estimate the relative magnitude of speckle fluctuations. The ratio $\sigma / \langle I \rangle$ in temporal statistics is analogous to the contrast of speckle patterns in spatial statistics.

If the integration time is large the high-frequency fluctuations are averaged while the low-frequency fluctuations still remain. Sometimes it is necessary maintain these conditions to take measurements, thus this high integration time can be simulate by adding repeated measurements many times. If the spatial standard deviation σ_s of the time-integrated dynamic speckle pattern is taken, the contrast $\sigma_s / \langle I \rangle$ becomes velocity dependent. Thus, the velocity distribution in the illuminated area can be observed as a contrast distribution.

Another approach to the subject is to analyze the time-differentiated intensity fluctuations. The ratio $\sigma_d/\langle I \rangle$ of the standard deviation of time-differentiated intensity fluctuations to the mean intensity gives a velocity-dependent value. If this ratio is taken at all points of the observed area using many frames of the speckle pattern, the result shows a velocity map. The statistics can be equivalent to those taken for the intensity difference between two successive time points, instead of the differentiated intensity. That is, they are equivalent to use the so-called structure function

$$s(\tau) = \langle [I(t) - I(t+\tau)]^2 \rangle \tag{4.3}$$

When the spatial standard deviation σ_{ds} of the time-differentiated biospeckle patterns is taken, the contrast $\sigma_{ds}/\langle I \rangle$ is also a function of the object velocity. The distribution of speckle moving (or changing) velocity is expressed by the contrast distribution.

4.4 SECOND-ORDER STATISTIC METHOD

The use of second-order temporal statistics is the most popular method for measuring the velocity or mobility of scatters in biological and nonbiological objects. Data is taken from devices like CCD cameras or photodetectors offering information in 2D or 1D dimension, respectively. In the bibliography section a wide variety of papers can be found reporting the time history speckle pattern, THSP[2], or Space Time Speckle, STS[3], which is the basis of second-order statistics such as autocorrelation, inertial moment, and statistical cummulants.

4.4.1 How THSP Is Formed

For every state of the phenomenon being assessed, successive images of speckle pattern are registered by means of a CCD camera and a frame grabber. Each image is digitalized to 8 bits gray levels, and just one column (e.g., the middle one) is stored in the frame grabber memory. Then a new pseudoimage is composed by setting each column side by side. A new image is conformed when the number of columns completes an image, providing a matrix with $m \times n$ values, where m represents the lines, or the time history, of each pixel of the column n taken from each speckle pattern. The new image offers information on the time evolution of the speckle pattern desired throughout each line, and is called the time history of speckle pattern (THSP). See Figure 4.3.

Here we should consider how the image of the speckle pattern is acquired. Figure 4.1 presents two ways to register the speckle pattern produced by an active sample with a CCD camera. In the first case (Figure 4.1a), the photo detector of the CCD camera registers the speckle pattern (SP) produced by free propagation of the wave front output from the sample, without a lens. Another way to obtain the SP image is imaging the sample on the detector by means of an optical system.

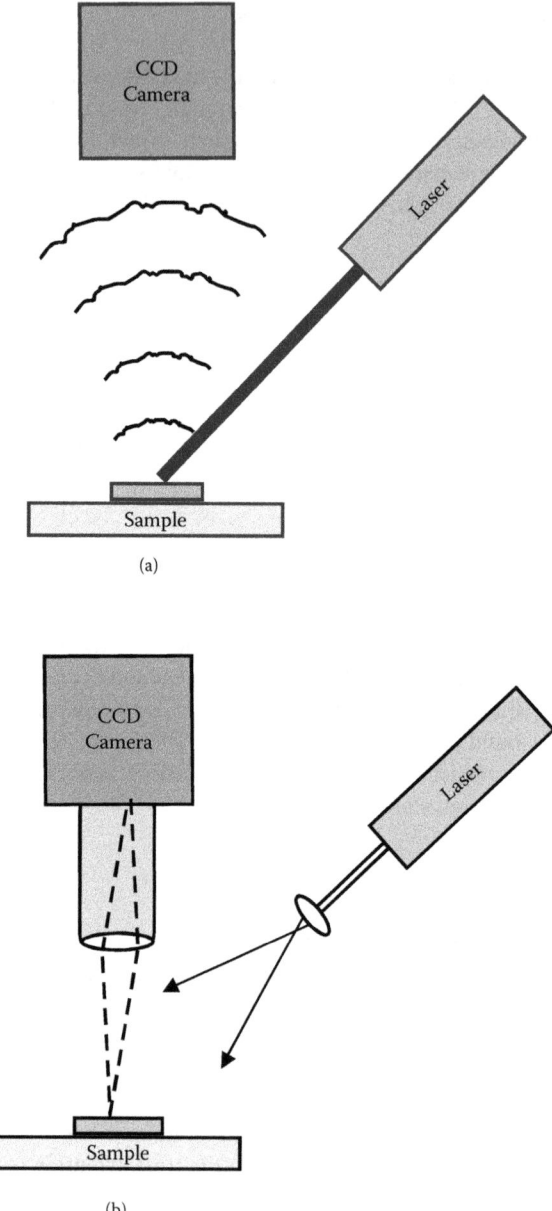

FIGURE 4.1 Speckle pattern (a) produced by free propagation of a wave front oncoming from the sample without a lens and (b) produced when the sample is imaged on the CCD photo detector through an optical system.

In free propagation, each point on the detector receives the contribution of all points in the illuminated sample. Otherwise, if the sample is imaged on the CCD photo detector through an optical system, each point from the sample contributes to only one point in the image, Figure 4.1b. If the optical system is defocused, each point in the observation plane receives information from an area in the sample whose size depends on how defocused it is.

4.5 AUTOCORRELATION FUNCTION

Correlation is a mathematical tool used frequently in signal processing for analyzing functions or series of values, such as time domain signals, providing a mutual relationship between two or more random variables. The autocorrelation function is defined as the cross-correlation of the signal itself. Autocorrelation is useful for finding repeating patterns in a signal, such as determining the presence of a periodic signal that has been buried under noise, or identifying the fundamental frequency of a signal which does not actually contain that frequency component but implies it with many harmonic frequencies.

There are different definitions of autocorrelation, but not all of them are equivalent. In some fields, the term is used interchangeably with autocovariance.

In statistics, given a stochastic process $f(t)$, the autocovariance is simply the covariance of the signal against a time-shifted version of itself.

In signal processing, given a signal $f(t)$, the continuous autocorrelation $Af(\tau)$ is the continuous cross correlation of $f(t)$ with itself, at lag τ, and is defined as:

$$Af(\tau) = f^*(-\tau) \otimes f(\tau) = \int_{-\infty}^{\infty} f(t+\tau)f^*(t)dt = \int_{-\infty}^{\infty} f(t)f^*(t-\tau)dt \qquad (4.4)$$

with the cross circle representing the convolution operation.

Some of the properties of the autocorrelation are:

- The continuous autocorrelation function reaches its peak at the origin, where it takes a real value.
- The autocorrelation of a periodic function is itself periodic with the very same period.
- The autocorrelation of the sum of two completely uncorrelated functions (the cross correlation is zero for all τ) is the sum of the autocorrelations of each function separately.
- Because autocorrelation is a specific type of cross correlation, it maintains all the properties of cross correlation.

An often-useful alternative for evaluating signals is the power spectral density (PSD), which describes how the power of a signal or time series is distributed with frequency. Here, power can be the actual physical power or, more often for convenience with abstract signals, it can be defined as the squared value of the signal, that is, as the actual power if the signal was a voltage applied to a 1 ohm load. This instantaneous power (the mean or expected value of which is the average power) is then given by:

$$P = |F(v)|^2 \tag{4.5}$$

The Wiener-Khinchine theorem states that the power spectral density of a wide sense-stationary random process is a Fourier transform of the corresponding auto-correlation. As set out, the equation 4.4 the autocorrelation is defined by:

$$Af(\tau) = \int_{-\infty}^{\infty} f(t+\tau)f^*(t)dt$$

On the other hand we know that by defining the Fourier transform

$$f(\tau) = \int_{-\infty}^{\infty} F(v)\cdot\exp(2\pi iv\tau)dv$$

where: $F(v) = \Im\{f(\tau)\}(v)$ is the Fourier transform of $f(\tau)$
Taking the complex conjugate

$$f^*(\tau) = \int_{-\infty}^{\infty} F^*(v)\cdot\exp(-2\pi iv\tau)dv$$

replacing in the equation of the autocorrelation

$$Af(\tau) = \int_{-\infty}^{\infty}\left[\left[\int_{-\infty}^{\infty} F^*(v)\cdot\exp(-2\pi iv\tau)dv\right]\left[\int_{-\infty}^{\infty} F(v)\cdot\exp(2\pi iv(t+\tau))dv\right]\right]dt$$

after algebra

$$Af(\tau) = \int_{-\infty}^{\infty} F^*(v)F(v)\exp(2\pi iv\tau)dv = \int_{-\infty}^{\infty} |F(v)|^2\exp(2\pi iv\tau)dv$$

$$Af(\tau) = \Im^{-1}\{|F(v)|^2\}(\tau) \tag{4.6}$$

So inversely the power spectral density is the

$$|F(v)|^2 = \Im\{Af(\tau)\}(v) \tag{4.7}$$

The Wiener-Khinchine theorem is a special case of the cross-correlation theorem were the cross-correlation is performed on the same function itself

The measuring techniques based on second-order temporal statistic are also known as dynamic light scattering spectroscopy, photon correlation spectroscopy, light beating spectroscopy, and intensity fluctuation spectroscopy. The autocorrelation function or the power spectrum density function of temporal intensity fluctuations of the speckle pattern are used in this approach. These functions directly reflect the frequency components of the speckle signals and the dynamics of the objects. In some cases, they are closely related to laser Doppler velocimetry. Fluctuations of speckle patterns can be understood as the result of a time-varying distribution of individual scatterers within a random structure of grains. Decorrelation techniques are useful in assessing the time dependence of biological samples.[4]

A measure of the mean lifetime of a speckle could be correlated to the dynamic properties of the biological activity. Full width at half maximum (FWHM) of the autocorrelation applied in a time history of the speckle pattern (THSP) has been used to measure speckle grains lifetime.[5,6] Such a measurement is somewhat noisy because it depends on only one value of the autocorrelation curve with the other points discarded. One possible form of reducing noise is considering several autocorrelations over all the rows of the THSP, averaging the width obtained at several fractions of their maximum heights,[2] being the widths of equivalent rectangles (WER).

Let suppose that the function we calculate the autocorrelation is a rectangle function. The normalized autocorrelation function is $y = 1 - \frac{x}{b}$ where b the width of the rectangle. The ordinate corresponding to half of maximum is $x' = b/2$. The same value of b can be obtained using different fractions of the height other than the half. Let us choose the fraction $1/n$ of the full height. If the function considered is a rectangle, the width of this fraction is

$$b_n = \frac{x_n}{1 - 1/n} \qquad (4.8)$$

where x_n is the abscissa of the point in the function that corresponds to $1/n$ of the maximum value.

In order to use more points of the autocorrelation data, the results obtained at several different fractions could be averaged. The mean value obtained in this way is

$$\langle b_n \rangle = (1/n) \sum_{n=1}^{N} \frac{x_n}{1 - A(x_n)} \qquad (4.9)$$

where N is the number of used data points. Therefore, the expression presented in Equation 4.9 is the definition of WER, and the integral mean value is defined as a rectangle to be applied in THSP signals. Nevertheless, observing a line profile of THSP the function can be approximated with other kind of curves with smooth behavior, such as a decreasing exponential decay, Gaussian, or Lorentzian function.[6] These functions have similar autocorrelations. Therefore, if the autocorrelation were adjusted by mean of these functions, its characteristic width would give us measurements of activity of the phenomenon that gives rise to the THSP.

An alternative way to adjust the autocorrelation function is by means of statistical cummulants. The statistical cummulants is an approach to obtain the coefficients of a polynomial regression applied to the autocorrelation function of the THSP. The normalized autocorrelation function may be adjusted by exponential decay, or like a Gaussian curve. Initially, the autocorrelation function can be considered statistically as a characteristic function[7] defined as:

$$G'(t) = \exp\left\{ \sum_{i=1}^{\infty} \frac{K_n}{n!} (it)^n \right\}$$

(4.10)

where K_n are the cummulants of the distribution. Even if it is not secure to consider the autocorrelation as a characteristic function of the intensity fluctuations, equation (4.10) is adequate for describing the autocorrelation function by considering only the first few points. The adjustment must be done by using a function such as:

$$G(\tau) = a + (1-a)\exp\left[-\left(b_1\tau + b_2\tau^2 + b_3\tau^3 + \cdots + b_n\tau^n \right) \right]$$

(4.11)

With b_i being associated with the cummulants as $b_i = \frac{K_i i^n}{n!}$

One example of the use of statistical cummulants was described by Rabelo et al.[15] The statistical cummulants were used as quality and senescence indicators for orange fruit.

Another approach would be to use all the data in the autocorrelation with the heuristic measure suggested by the concept of entropy. It is not itself an entropy, because there is not a condition on the sum of all autocorrelation values equivalent to the normalization of a probability density. The autocorrelation values $A(x)$ are mapped as:

$$M = \sum A(x)\log A(x)$$

(4.12)

where the sum is taken over all the experimental values of the autocorrelation excluding $A(0)$. The number thus obtained is considered to be related to the graininess of the THSP, which is expected to decrease as the lifetime of those speckled increases.

For testing the relative performances of different methods, computer simulations were implemented. A set of noisy simple profiles were used to conform an image of a THSP as simulated samples. Mean values of the FWHM, WER and X*LOGX were calculated and standard errors were plotted against noise amplitude. The results suggest greater resolution and good repeatability, particularly using the WER method, which shows better behavior with respect to noise tolerance. Figure 4.2 shows relative errors in percents obtained by three measurements: FWHM, X*LOGX, and WER.

4.5.1   EXAMPLE OF A PAINT DRYING PROCESS

In order to illustrate the use of the approaches presented, one actual example is shown. It is a paint drying process under laser illumination. The samples were illuminated with a low power laser beam attenuated to avoid any influence of the radiation on the

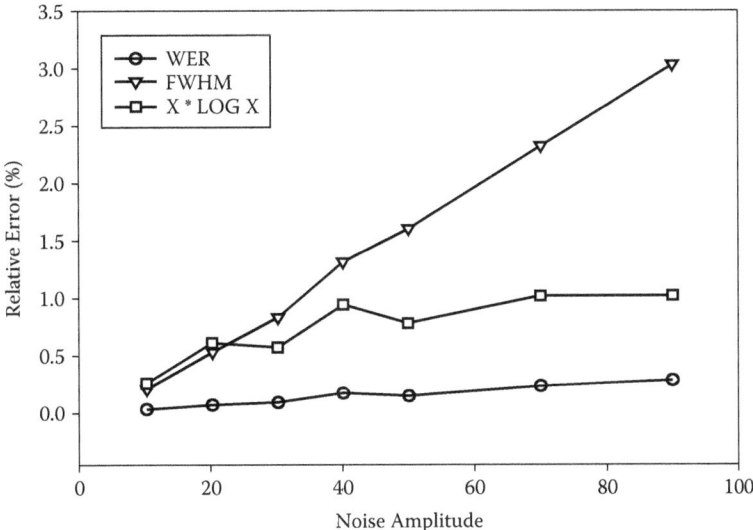

FIGURE 4.2 Relative error in percents in a computer simulation of three types of measurements FWHM, X*LOG X, and WER. (Figure with permission from Rabal, H. et al. Transient phenomena analysis using dynamic speckle patterns, *Opt. Eng.* (1996), 35(1), 57–62, SPIE).

specimen, and successive images of the free propagation wave front were registered by a CCD camera and a frame grabber. The THSP images were obtained every 2 minutes, so the composite image shows the time evolution of the speckle pattern. A flat glass plate was coated with a thin film of enamel paint and speckle activity was recorded during the natural process of drying.

When a phenomenon shows low activity, time variations of the speckle pattern are slow, and the THSP shows elongated shape. When the phenomenon is very active, the THSP resembles an ordinary speckle pattern. The mere look of the THSP image gives an idea of how active is the sample. Figure 4.3 shows the THSP of two possible states of the drying of paint and one typical line for each that shows both profiles.

Figure 4.4 shows the autocorrelation functions for different states of drying, with the abscissa units corresponding to the time that the image processor requires to construct a THSP image. As drying proceeds, the result tends to correspond to the autocorrelation of a rectangle function, thus showing the growing lifetime of the speckles. As the first few points of the autocorrelation show greater differences in the graphic, only the 30 first values are represented.

Figure 4.5 shows the results of applying FWHM, WER, and X*LOG X to the drying of paint. It is easy to see that both WER and X*LOG X give better information about the drying process than FWHM. The greatest separation between the drying curves occurs on the first 30 points. It was the reason why only those points were used to calculate the WER, thus improving the discrimination.

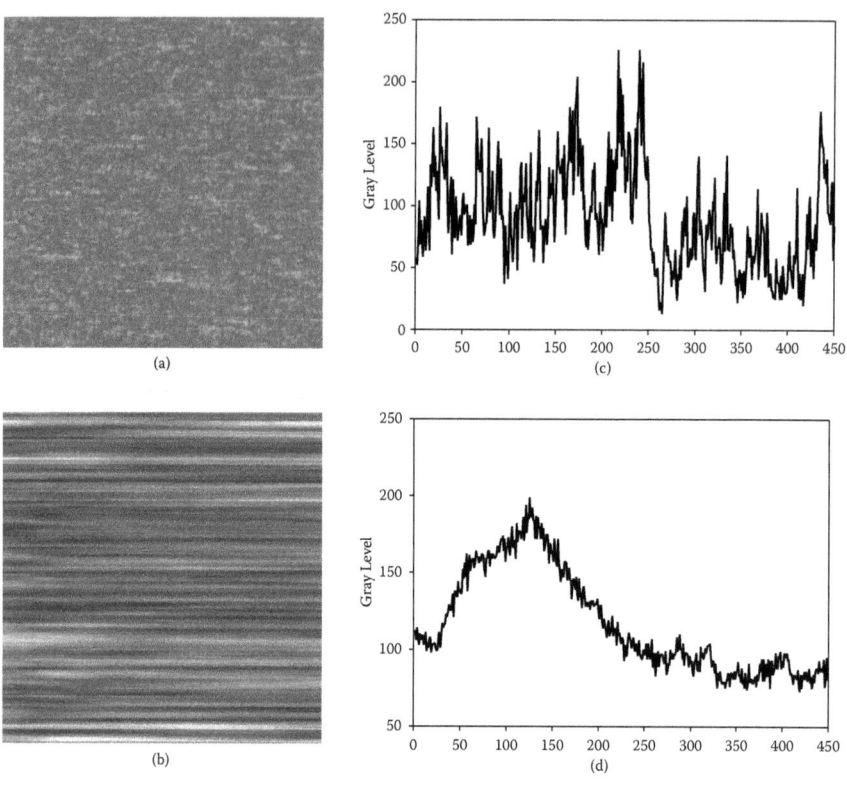

FIGURE 4.3 THSPs of two different states of drying of paint and their profile lines, respectively.

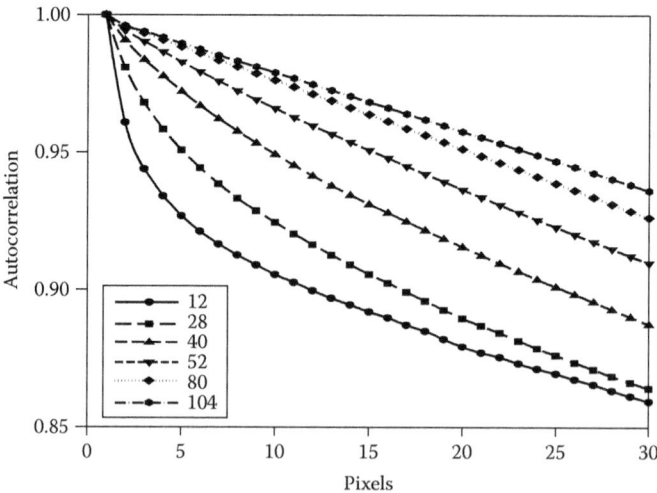

FIGURE 4.4 Autocorrelation functions of the THSP corresponding to a drying of paint, with the parameters indicating the drying time in minutes.

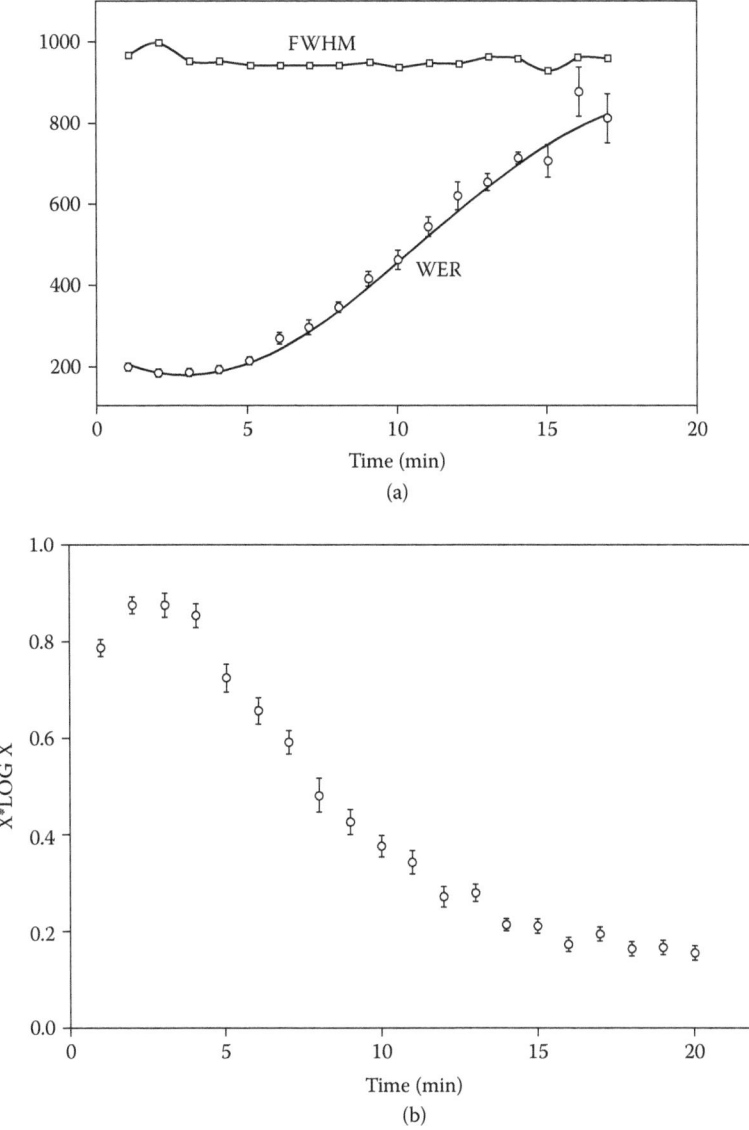

FIGURE 4.5 Curves of (a) FWHM and WER and (b) X*LOG X as function of drying time of paint. (From Rabal et al., *Opt. Eng.* (1996), 35(1), pp. 57–62.)

4.6 INERTIA MOMENT METHOD

An alternative way to characterize speckle time evolution is based on the cooccurrence matrix[8–10] of the intensity in the time domain.

Cooccurrence matrix (COM) is defined as

$$COM = [N_{ij}] \qquad (4.13)$$

The entries are the number N of occurrences of a certain intensity value i that is immediately followed by an intensity value j. This is a particular case of the so-called "spatial gray level dependence matrices." It is usually used to characterize texture in images.[11] In the spatial case, its principal diagonal is related to homogeneous regions, and the nonzero elements far from it represent high-contrast occurrences.

In our case, the variable of interest is time, then the involved N values are the occurrences of a certain gray value i followed in the next time step by a value j in the THSP as described previously.

When the intensity does not change, the only nonzero values of this matrix belong to its principal diagonal. As the sample shows activity, intensity values change in time, the number N outside the diagonal increases, and the matrix resembles a cloud. For normalization purposes, it is convenient to divide each row of this matrix by the number of times that the first gray level appears.

$$M_{ij} = \frac{N_{ij}}{\sum_j N_{ij}} \tag{4.14}$$

Then, the sum of the components in each row equals 1.

This modified cooccurrence matrix (MCOM) is an experimental approximation to the transition probability matrix between intensity values in the THSP. It corresponds to a generalization of the histogram and bears a certain resemblance to the transition matrix appearing in the Markov processes.

Figure 4.6 shows the two situations—(a) high and (b) low activity—cooccurrence matrices, corresponding to the two examples shown in Figure 4.3 and representing a wet and an almost dry paint. It can be seen that when the phenomenon is very active, the associated cooccurrence matrix is spread; otherwise, it is concentrated around the principal diagonal.

A measurement of the spread of the M values around the principal diagonal with these features can be constructed as the sum of the matrix values times its squared row distance to the principal diagonal. This is a particular second order moment called the *inertia moment* (IM) of the matrix, with respect to its principal diagonal in the row direction. The name is suggested by the mechanical analogue of this operation.

So, inertia moment (IM) is defined as

$$IM = \sum_{ij} M_{ij}(i - j)^2 \tag{4.15}$$

This measurement is similar to one currently used in photon correlation spectroscopy, called *photon structure function*.[12]

This measurement (IM) is a useful tool to estimate the global activity in the several biological and nonbiological applications with a summary value.

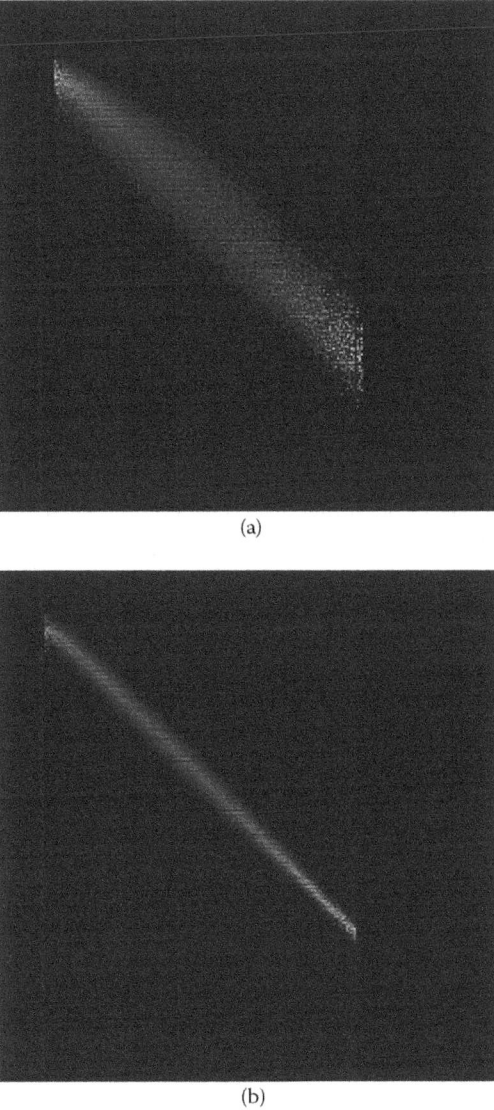

FIGURE 4.6 Cooccurrence matrices for two stages in paint drying, related to the THSPs presented in this chapter1. (a) High activity. (b) Low activity.

This measurement also includes as a summary contributions of all the process present in the sample. A single measure could never give complete information about the phenomenon source that produces the dynamic speckle. It is not enough to correlate the activity with any characteristic of the sample to be investigated. Another parameter commonly used in the literature to characterize the activity of a sample is the Briers[13] contrast (BC). He defined speckle contrast in time direction as the ratio of the time standard deviation to the mean intensity value. He has shown that temporal contrast is related to the proportion of intensity scattered by moving scatters to the total intensity by mean of the following expression:

$$\frac{\left\langle \sigma_t^2(x,y) \right\rangle_{x,y}}{\left\langle I \right\rangle^2} = 1 - (1 - \rho^2)$$

where $\left\langle \sigma_t^2(x,y) \right\rangle_{x,y}$ is the special mean value of the time variance of the intensity in pixel (x,y), $\left\langle I \right\rangle^2$ is the mean value of the intensity (space and temporal), and r is the ratio of the intensity scattered by moving centers and the total intensity. Thus it is possible to analyze as scatters will be immobilizing as the paint dries. e.g. the evolution of the drying paint process.

In some cases it is not necessary to use the whole THSP to calculate IM or BC, allowing the possibility of obtaining just a portion of it, such as only some rows, or even a short number of pixels of the whole image, providing a reliable result.

An analysis of the reliability of the THSP images for different simulated and experimental situations was carried out. Brier's contrast (BC) and inertia moment (IM) of the cooccurrence matrix were used to analyze the influence of the shape and size of different windows. Both square and rectangular windows were tested in the space–time domain, and the impairment of the results was analyzed as the number of the data was diminished. Figure 4.7 shows both BC and IM, calculated for square and rectangular windows.[14]

The reduction of the window size results in some impairment of both measures. Nevertheless, even if the error is high for small windows, the comparison between two states, as is the case of dead and alive seeds, can be reliably performed with small windows.

An additional question that must be taken into account in the construction of THSP images is the intensity level of the detector. When the THSPs are being recording, it is necessary to avoid exceeding the saturation value of the CCD detector, because if the variations surpass this value they will be unnoticed. Something similar happens for lower values of intensity fluctuations; if the variations are smaller than the quantization interval, the system is unable to give information about them. All these reasons contribute to the inaccuracy in the IM measurements. One example can be noticed if you add or subtract a constant to all intensities of a THSP. It should not change the IM value unless saturation or underexposure occurs.

Figure 4.8a represents a THSP line profile, and the same one with new values, after adding and subtracting a constant to the gray values from the original image, are represented in Figure 4.8b,c, respectively. It is easy to see that in these two signals, part of the information has been lost. The IM values will be different due to this loss of information. So it is possible to conclude that some careful adjustment could be made, and standardization should be done to avoid those effects.

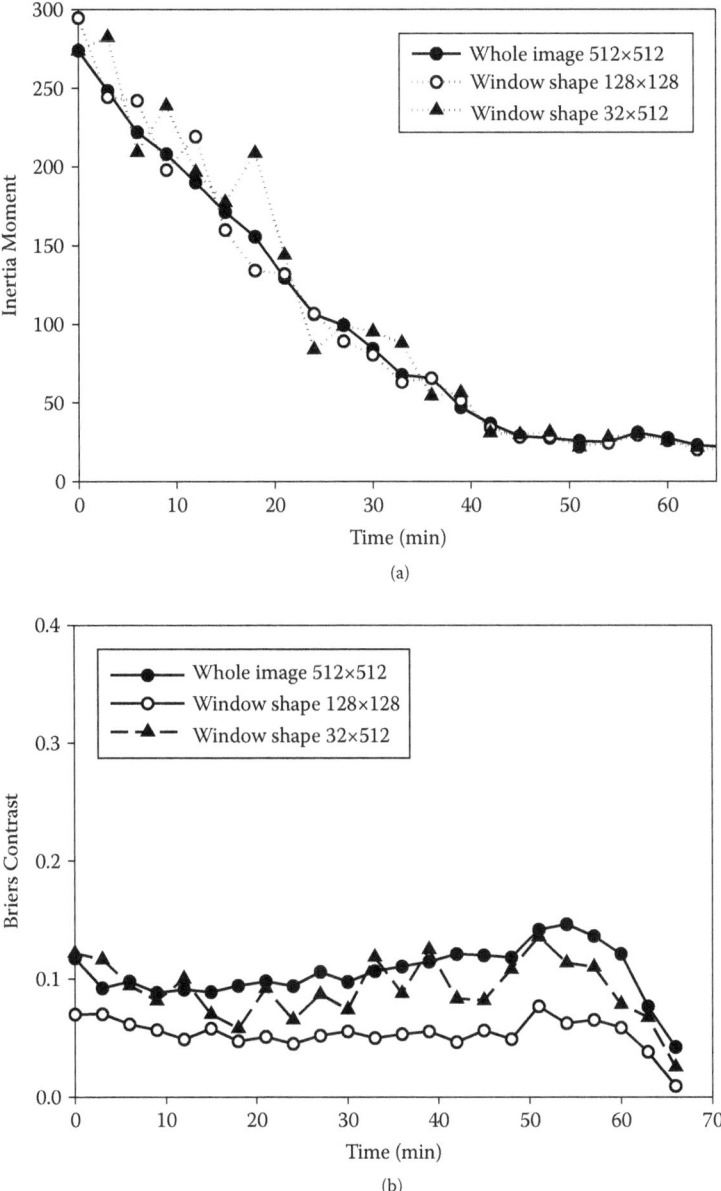

FIGURE 4.7 Drying in paint drying experiment for two different shapes of smaller windows: (a) inertia moment, (b) Brier's contrast. Both are plotted against time.

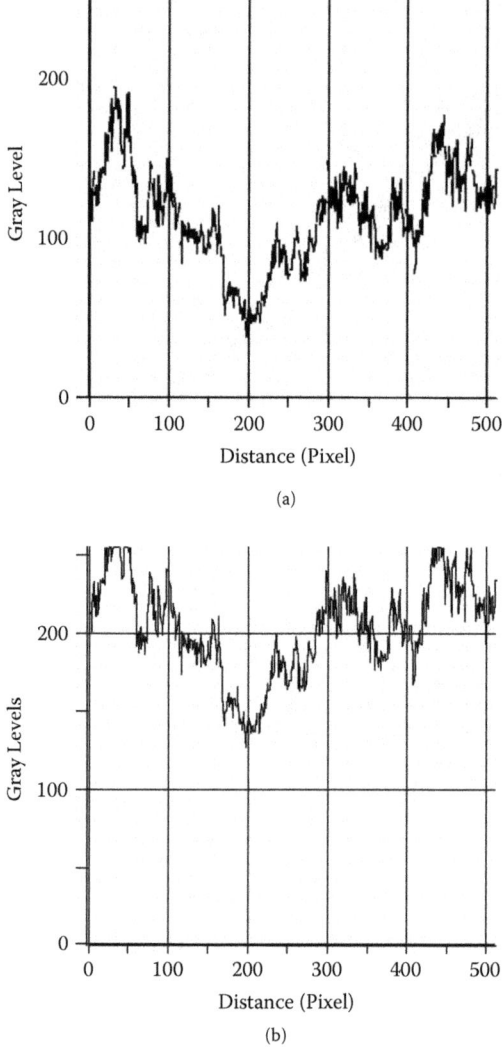

FIGURE 4.8 Line profile of (a) THSP, (b) THSP with a constant added, and (c) THSP with a constant subtracted.

THSP images are different for each application, and IM results are directly related to all line profiles. Equation 4.10 indicates that the main information in the IM calculation is the difference between two immediate pixels, so, if the time history is characterized by a low frequency, modulating the curve, that low frequency behavior will not be represented in the IM calculus.

Other aspects that must be taken into account for the measurement of the IM will be exemplified with a different frequency and amplitude composition in the time history of a THSP image.

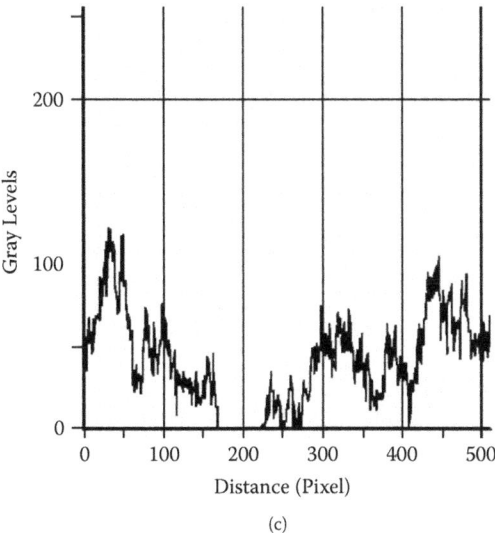

(c)

FIGURE 4.8 (*Continued*)

Let us suppose three different activities characterized by the line profiles shown in Figure 4.9. It could be supposed that the last curve should have a low IM value because the amplitude of the variation is lesser than the other ones. However, high amplitude values do not necessarily mean high values IM, but the fast variations are responsible to raise the value of MI. In Figure 4.9 it is exemplified with respective IM values representing the major influence of high frequencies in IM values.

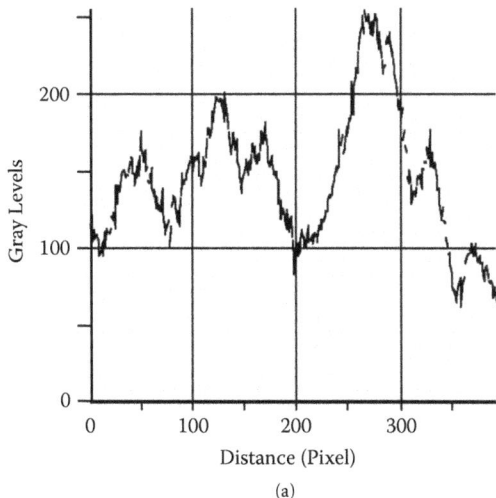

(a)

FIGURE 4.9 The graphic shows three line profiles with different IM values. Notice the major influence of high frequencies on IM values: (a) IM = 25.8, (b) IM = 35.6, and (c) IM = 292.3.

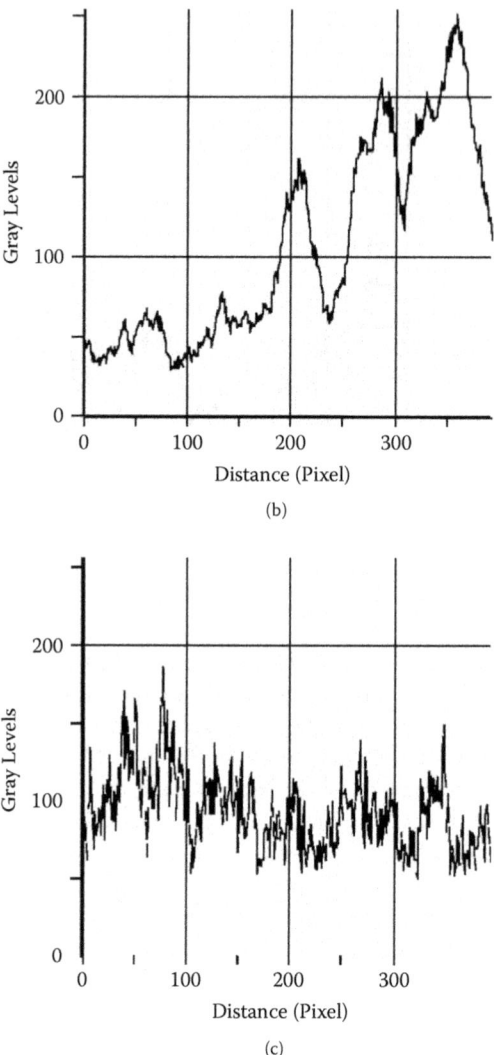

FIGURE 4.9 (*Continued*)

4.7   CONCLUSIONS

We have analyzed many of the technical problems that must be considered in any dynamic speckle experiment from the point of view of statistical analysis. The phenomena can be characterized by space–time correlation functions and power spectrum of the speckle intensity fluctuations, which depend on the illumination light and the optical configurations.

   The study of dynamic speckle observed in light fields scattered from living objects has drawn special attention from many investigators when termed *biospeckle*.

Nevertheless, the results described in this chapter are also important for applications to speckle metrology, either in biological or nonbiological specimens.

There are, though, many other branches of scattering phenomena that have close parallels with the dynamic speckle discussed here. However, this could be a starting point for developing other applications.

In conclusion, we hope that the dynamic speckle phenomena will be further investigated for several kinds of phenomena and applied to measurement of these various physical or biological behaviors.

REFERENCES

1. Briers, J. D., Wavelength dependence of intensity fluctuations in laser speckle patterns from biological specimens, *Opt. Comm.,* 13, 324, 1975.
2. Rabal, H. J. et al., Transient phenomena analysis using dynamic speckle patterns, *Opt. Eng.*, 35(1), 57, 1996.
3. Xu, Z., Joenathan, C., and Khorana, B. M., Temporal and spatial properties of the timevarying speckles of botanical specimens, *Opt. Eng.*, 34(5), 1487, 1995.
4. Oulamara, A., Tribillon, G., and Duvernoy, J., Biological activity measurement on botanical specimen surfaces using temporal decorrelation effect of laser speckle, *J. Mod. Opt.*, 36(2), 165, 1989.
5. Tanaka, S., Takenaka, I., and Ohtsuka, Y., Statistical evaluation of phase fluctuations of light scattered from apple peel, in *Opt. Meth. in Biomed. and Environ. Sci.*, Ohtzu, H. and Komatsu, S., Eds., Elsevier, Amsterdam, 1994, 15.
6. H. Cummins and H. Swiney, Light beating spectroscopy, in *Progress in Optics*, Wolf, E., Ed, SPIE, Amsterdam,1970, 8, 133.
7. Chu, B., *Laser Light Scattering*, Academic Press, New York, 1991, chap. 7.
8. Haberacker, P., *Digitale Bildverarbeitung Grundlagen und Anwendungen*, Hanser, C., Ed., Verlag, München, 1985.
9. Zobrist, A. and Thompson, W., Building a distance function for gestalt grouping, *IEEE Transac. Comp.*, 24(4), 718, 1975.
10. Kruger, R., Thompson, W., and Turner, A. F. Computer diagnosis of pneumoconiosis, *IEEE Transac. Sys., Man Cybern.*, 4, 40, 1974.
11. Allam, S., Adel, M., and Réfrégier, P. Fast algorithm for texture discrimination by using a separable ortonormal decomposition of the co-occurrence matrix, *Appl. Opt.*, 36, 8313, 1997.
12. Chu, B., *Laser Light Scattering: Basic Principles and Practice*, Academic Press, Boston, 1991, chap. 4.
13. Briers, J. D., The statistics of fluctuating speckle patterns produced by a mixture of moving and stationery scatterers *Opt. Quantum Electron*, 10, 364–366, 1978.
14. Braga, R. A., Silva B.O., Rabelo, G., Costa R. M., Enesa, A. M., Cap, N., Rabel, H., Arizaga, R., Trivi, M., and Horgand, G., Reliability of biospeckle image analysis, *Opt. Lasers Eng.*, 45, 390, 2007.
15. Rabelo, G., Dal Fabro, I., Braga, R., Rabal, H., Arizaga, R., and Trivi, M., Laser speckle techniques in quality evaluation of orange fruits. *Revista Brasileira de Engenharia, Agricola e Ambiental*, 9, 570, 2005.

5 Activity Images
Generalized Differences, Fujii´s, LASCA, and Related Methods

Hector Jorge Rabal

CONTENTS

5.1   ACTIVITY IMAGES: GENERALITIES

In this chapter we are going to describe some features related to activity images in dynamic surfaces, mainly those obtained by using Generalized Differences, the LASCA method, and the method suggested by Fujii et al. Other algorithms will be discussed in other chapters. Mathematic details are not going to be described in depth as they can be found in the bibliography.

In many actual experimental situations the speckle activity shown in the image of an object is not uniform over the surface of the sample. It may show local differences due to variations in its many possible origins. In biological samples, for example, local variations may be due to normal ("healthy") biological reasons, its intrinsic variability, or due to pathological reasons—increased or diminished activity due to bruising, presence of pathogens, enhanced or inhibited blood flow,[1] and many others.

Activity images are representations where some features of dynamic speckle is depicted as grey levels or colors. Blood flow is one of the phenomena most extensively explored and commercially available instruments already exist for obtaining perfusion images.[1-2]

Diabetes, arteriosclerosis, and vasospastic conditions (sudden constriction of a blood vessel, causing a reduction in blood flow) associated with Raynaud's syndrome, are a few of the abnormalities that can happen in or near human skin and that could be screened by using some of these techniques. Exposed inner regions, such as in wounds or burns, or some others by use of endoscopes can also be accessed.

The screening of such variations can, in some cases, be used as a diagnostic tool of its origins or evolution under treatment, as is the case of the presence of parasites, for example. The use of activity depiction in images is mostly a qualitative tool, although a few quantitative uses have been reported.

Several algorithms have been proposed for the generation of these activity images. LASCA, averaged differences, generalized differences, Hurst coefficient, spectral bands, blur coefficient, statistical cumulants, temporal contrast and entropy, are some of them (see references). Some of these are heuristic, and a few are based on physical considerations requiring the fulfillment of some assumptions and some knowledge of the investigated phenomenon. All of them require a different kind of interpretation of the results. There are no available dynamic speckle standards to permit a reliable calibration or even standardization.

The only (nonoptimal) approaches are the use of numerical simulations with models and/or the help of experts in the identification of the causes by alternative methods. Rotating[3] and linear[4] moving diffusers have been used as variable controlled sources for calibration. Those are repeatable but only simulate one feature of the possible origins of speckle activity.

A comparison of two states with different activity before and after some activation, in a similar way to functional magnetic resonance images, is an alternative easier to implement. A comparison to the baseline is made through a ratio or a difference between images under the variation of some external parameter.

Some techniques can be implemented in real time or in almost real time, with either scanning or no scanning set-ups, in noninvasive ways. Advantages to alternative methods are the noncontamination, nonperturbative character of the measurements and the relatively low cost of the instrumentation. As activity depends on several possible origins, activity images also will depict different features depending on the used algorithms and registering methods. Interpretation of the results is then, in many cases, heuristic, and may depend heavily on the experts' knowledge. Nevertheless, some techniques admit, at least partially, a physically based interpretation. The screening of regions showing different activity may prove useful even in absence of a physical understanding of the origins of the differences.

When the approaches are heuristic, they are built under the proviso that the observations are indications that the same or similar reasons are actuating in the regions of the sample exhibiting the same value. This proviso may, or may not, prove to be reliable. As dynamic scattering is a "many to one" phenomenon, the same results may be originated by different sets of experimental situations.

When a series of results concerning the same samples under time evolving phenomena, such as for example drying, it is convenient to use the same dynamic range to show all the results, as otherwise some results would be boosted or compressed when compared with others. In addition, when the algorithm involves real numbers, it is also convenient to preserve them until the last operation of quantization before the image representation.

The use of activity images requires the assistance of an expert in the investigated effect to look for the adequate interpretation. Inhibited or exacerbated activity may be indications of pathologies. For example, the regions of a seed that show high activity may be indications of its vigor, or viability, or the presence of fungi.

Only phenomena occurring in the surface, or near it, or through transparent regions (as the case of the eye) can be accessed in most cases. Access to deeper regions is hampered by multiple scattering and absorption. These phenomena are governed by different dynamics and require different treatments to those described in this chapter.

Activity is associated to intensity changes in the history each pixel in the image. So, descriptors that measure such changes are candidates for screening the loci of equal activity regions.

5.2 ALGORITHMS OF BIOSPECKLE'S IMAGE ANALYSIS

We are now going to describe the different algorithms. To demonstrate the results that are obtained using them we need more than one kind of sample, as not all methods are appropriate for the same phenomenon.

Thus, we are going to use three samples usually used for study cases: the drying of paint over an object of known topography (a coin; see Figure 5.1), a corn seed in the beginning of its germination process, and an image of a bruised apple where the bruise cannot be perceived.

In the painted coin, it is expected that the paint will be less active where the topography leaves a thinner layer than in regions where the surface is shallow where the layer is thicker. In the corn seed, the endosperm is expected to show less activity

FIGURE 5.1 A coin before the application of paint.

than the embryo, and in the bruised apple, the bruised region is expected to show different activity from the healthy regions.

The most immediate candidate for depicting activity seems to be the subtraction of consecutive frames. Pixels where the intensity does not change will appear dark, whereas those that suffered changes will appear proportionally bright. This operation can be implemented in real time, and is usually adopted to indicate whether the phenomenon that is investigated is present.

This representation depends strongly on the sampling time, and will show changes only if they occurred during the interframe lapse. Besides, as dynamic speckle is a statistical phenomenon, the resulting image will show a speckled appearance. To reduce that inconvenience, the averaging of several instances may be used. Stationarity must then be assumed, and acquisition and processing times increased. The absolute value, or the square, must then be included to take account of variations in both senses.

We are going to show some examples of the cited operations in an experiment where an object (a coin) was covered with a layer of paint and illuminated by laser. The speckle patterns were recorded, and no hint of the topography of the object could be perceived. (See Figure 5.2.) Figure 5.3a shows the subtraction of two consecutive images. Loci of equal thickness are in similar stages of drying and show similar activities. If a region is drier, then the speckles in consecutive images do not change as much as in wet ones. So, they appear darker. This result shows a very speckled appearance.

Figure 5.3b shows the average of 400 subtractions, such as that in Figure 5.3a, revealing the underlying profile. Notice the improvement in spatial resolution. Speckle noise is substantially reduced in the final image.

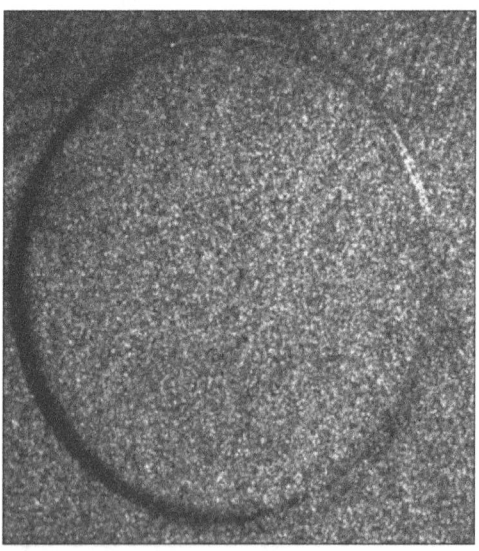

FIGURE 5.2 A speckled image.

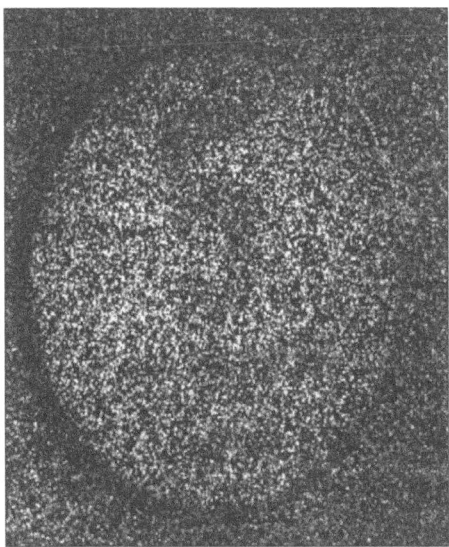

FIGURE 5.3a Subtraction of two consecutive frames.

Another very simple method for showing activity images uses the dynamic range[*] of the time series of each pixel.[5] Figure 5.4 presents an example that will be thoroughly discussed in coming chapters, consisting of a corn seed in the germination process under laser illumination and presenting two different regions: the endosperm and the embryo. These regions are different with respect to their biological activity,

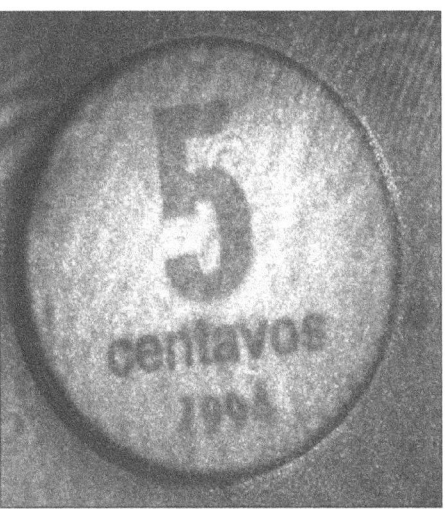

FIGURE 5.3b Average of 400 subtractions.

[*] Dynamic range is the difference between the maximum and the minimum values attained by the intensity in a pixel along time.

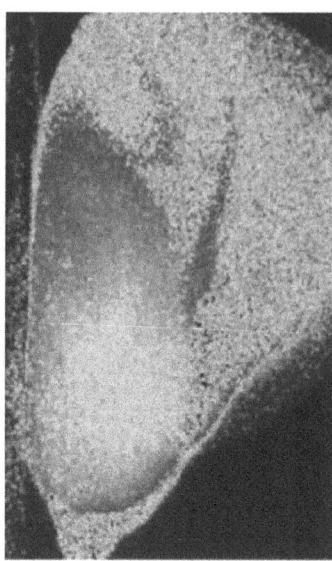

FIGURE 5.4 Dynamic range.

where the embryo is that which gives rise to life, and the endosperm is the source of embryo's food, therefore without live cells.

When a series of images of a time-evolving phenomenon is to be evaluated, and following Wardell,[6] two modes of presentation can be distinguished: a *relative* one, where the dynamic range of each processed image is expanded to occupy the full available range, and an *absolute* mode where the maximum and minimum values are chosen from the whole series to permit comparison between different images.

Techniques such as epiluminiscence (surface index matching with oil or other liquids) and nailfold capillaroscopy can be used as aids. Brewster angle illumination with polarized light can be used to benefit the proportion of light scattered under the surface.

5.3 GENERALIZED DIFFERENCES

A first and fast estimation of the activity distribution and the adequate sampling rate can be obtained by real time subtraction of an initial state to the following ones, as is usual in digital speckle pattern interferometry. This operation can be easily implemented in most image processors.

Inactive regions or low-activity ones can be expected to appear dark in these images. Conversely, regions presenting activities with typical evolution times comparable to the sampling rate can be expected to appear comparatively bright. Figure 5.3a shows a result obtained in this way.

As time between the current state and the reference one increases, speckles decorrelate, and the result very quickly deteriorates. Also, comparisons between images differing in short times (a few frames) are frequently not reliable. Resolution is low as images are speckled.

A first step to improving the result can be obtained by accumulating the (absolute value) subtraction of consecutive images in an operation that is equivalent to averaging. This operation, although requiring longer acquisition time, improves the results, as can be seen in Figure 5.3b.

This operation, weighted by the intensity average in the consecutive frames, is the algorithm used in Fujii's method and will be discussed later.

When the sample includes regions with different phenomena, which is both interesting and mostly the rule, the time scales of the intensity variations may differ. Then, it may be more convenient to include differences between nonconsecutive frames in the calculation.

This approach leads to the Generalized Differences (GD) idea.[7] In it, an image $I'_k(i,j)$ in constructed from a stack of consecutive frames by calculating the result given by Equation 5.1.

$$I'(i,j) = \sum_k \sum_l |I_k(i,j) - I_{k+l}(i,j)| \tag{5.1}$$

where $I_k(i,j)$ is the intensity at the point with coordinates (i, j) in the k-th frame, and the bars indicate absolute value.

That is to say, all the possible differences between all the different frames are added in absolute value for each point in the image. It means that the comparison between images is performed at all available time scales, and the results averaged. Impulsive noise, within the duration of a single frame, is blurred in this way and slower variations, requiring several frames to develop, are now included. As the $I_k(i, j)$ values are all compared, the GD result does not depend on the order of appearance of the values.

The GD operation is a minimum (zero) when and only when all $I_k(i, j)$ values are equal—in other words, when the pixel corresponds to a nonactive point in the sample.

It can be shown that the GD value is a maximum when the $I_k(i, j)$ values in the time histogram are evenly distributed near the bounds of the dynamic range. This is so when the lower value occurs half of the time and the other half the higher value appears. In actual experiments, it is convenient to avoid the higher end of the dynamic range of the frame grabber as it could hide higher intensity variations.

Then, if the time histogram of the intensity of a pixel shows two (or more) modes, the value of the GD is higher when the modes are more separated.

A heuristic condition can be included into the algorithm to discard variations occurring few times. Figure 5.5 shows a result obtained using this algorithm. The resulting image shows a noticeable reduction in speckle noise as compared with the formers.

Despite the reduction of speckle in image, the contours cannot be perceived with this technique. Nevertheless, this algorithm can be useful in seeds (corn, fungi, soy beans, etc.), bruising in fruits, and bacteria, which are described in the following chapters.

5.4 THE WEIGHTED GENERALIZED DIFFERENCES (WGD)

There are cases when the comparisons give better results when they are applied to certain subsets of the stack of recorded frames. When the phenomenon that is looked for is a fast one, requiring a few frames to develop, it is better depicted when the

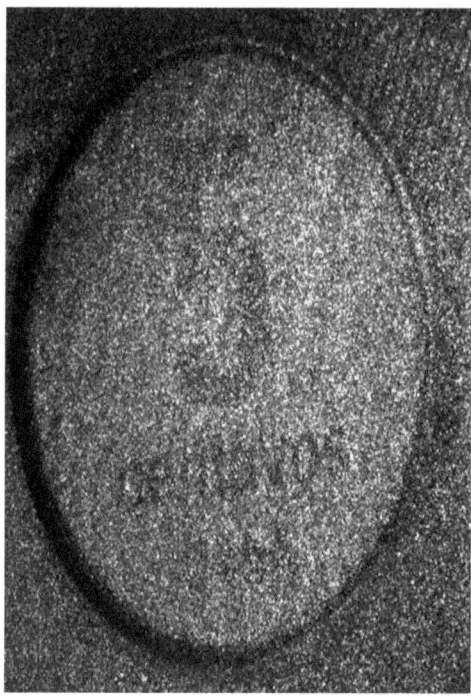

FIGURE 5.5 Generalized Differences algorithm applied to the drying of paint on a coin.

generalized differences are restricted to neighbors or close frames. Conversely, when the phenomenon requires several frames to develop, it is more convenient to subtract intensities only after a larger number of time steps. This is the case, for example, in some corrosion experiments.

To include the freedom of restricting the sub set of measurements that participates in the algorithm, the Weighted Generalized Differences[7] (WGD) is defined by Equation 5.2.

$$I'(i,j) = \sum_k \sum_l |I_k(i,\,j) - I_{k+l}(i,\,j)| p_l \qquad (5.2)$$

where the p values can be thought of as weights or running windows that are 1 (or other distributions) inside the subset of interest and 0 elsewhere.

It is not a priori obvious how the weights must be chosen to obtain the best results. An educated guess can be done by using the Receiving Operator Characteristic (ROC) curves.[8] The ROC method is a statistical instrument for taking rational decisions in tests with dycotomic results requiring a threshold.

The true value of the tested samples is assumed to be known, so that the fraction of times that the test identifies correctly a condition, called the TPF (True Positive Fraction or Sensitivity S), and the fraction of times identifying that the condition is not fulfilled, called the TNF (True Negative Fraction or Specificity), can be calculated.

The plot of TPF versus (1-TNF) for all possible thresholds is the ROC curve. It gives the probability of correctly classifying a sample whose real state is positive with respect to the condition under study. When the test does not discriminate for the condition between the results it consists in a straight diagonal line. As the accuracy of a test increases, the curve displaces from the diagonal to the upper left corner and, if discrimination were perfect, it would contain that point.

These curves are a good index of the discrimination capacity between alternative states, and so it can be used to test for the best assignation of weights in WDG.

In this approach, when the aim is to screen regions involving a binary decision (healthy or sick, bruised or not bruised), the use of ROC curves is useful to test several weights distributions and to compare the ROC curves to determine which is the best choice for the required purpose.[8] This analysis must be performed for each type of experiment. As can be seen in Figure 5.6a, the WGD for that experiment provides better discrimination (the curve is higher for all the abscissa) when the weights are located in the nearest 5 neighbors.

The edges and the coins inscriptions can be perceived so that the WGD performs as a better alternative to GD in applications such as this one (see Figure 5.6b).

For applications, both GD and WGD methods suffer from the shortcoming of requiring many images, so the investigated phenomenon must be slow as compared with the sampling rate and stationary during image acquisition periods. Corrosion, slow drying of coatings, low activity in seeds, efflorescence, etc., are some possible candidates.

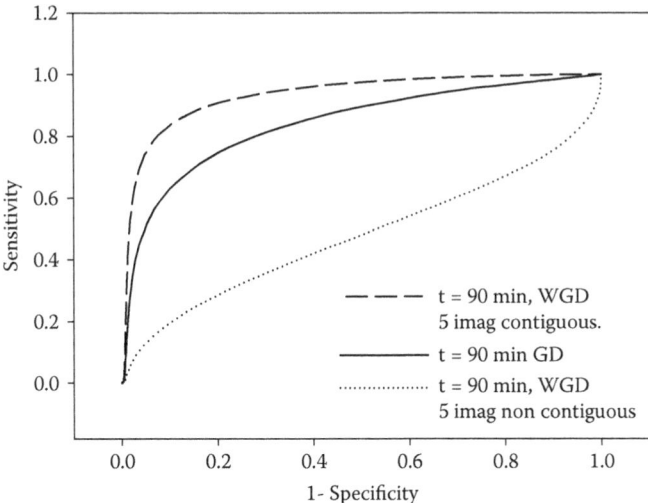

FIGURE 5.6a ROC curves for the Weighted Generalized Differences: Nonzero weights in every five contiguous images, Generalized (nonweighted) Differences and nonzero weights in five noncontiguous images. (Reprinted from Proc. SPIE Vol. 6341, 63412B, Hector Rabal, Christian Ortiz, Marcelo Trivi and Ricardo Arizaga, Activity speckle measurement comparison using R. O. C. (receiver operating characteristic) methods, *Speckle 06: Speckles, From Grains to Flowers*, 2006.)

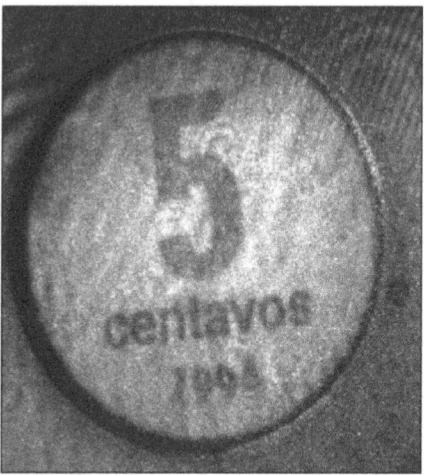

FIGURE 5.6b WGD applied to the drying of paint on a coin. Nonzero weights in only 5 nearest neighbors.

In order to test if WGD were adequate estimators of activity, we used the dynamic speckle obtained from a slowly rotating diffuser where the linear speed could be chosen by illuminated different radial regions. The experiment was as shown in Figure 5.7. A diffuser disk was rotated by a low-speed motor driver ($\omega = 0.1$ and 0.03 s^{-1}). Light from a He–Ne laser was scattered in a small region of the disk and registered by a photomultiplier tube (PMT).

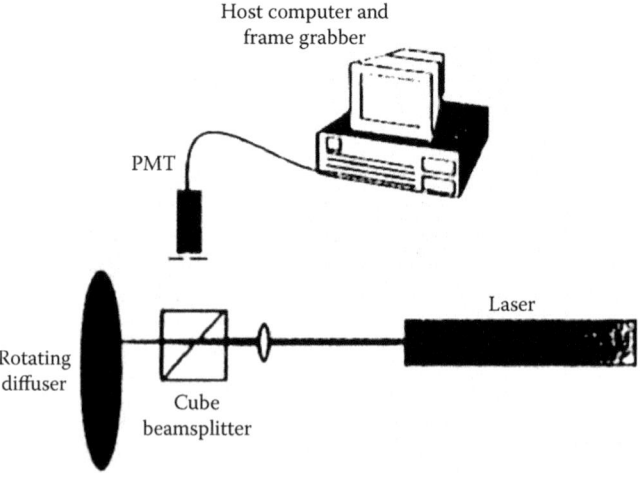

FIGURE 5.7 Simulation of a dynamic speckle: The experimental set up for the rotating diffuser experiment. (From Romero, G.G., Alanís, E.E., and Rabal, H.J., Opt. Eng., 39(6), 1652, 2000.)

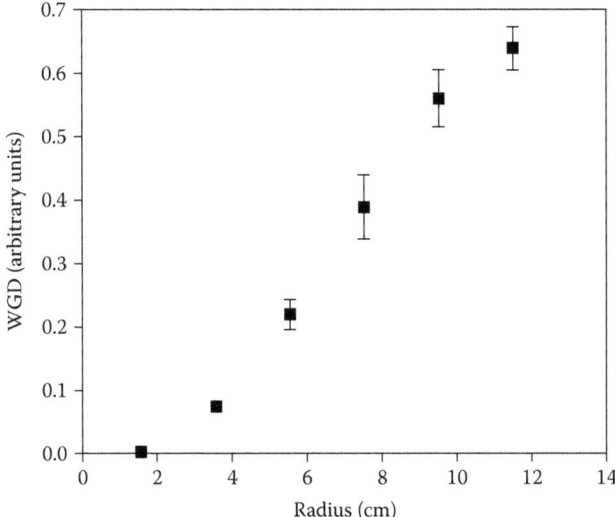

FIGURE 5.8 Weighted Generalized Differences for the dynamic speckle generated by a rotating diffuser as a function of the radius of the illuminated point on the diffuser. (From Romero, G.G., Alanís, E.E., and Rabal, H.J., Opt. Eng., 39(6), 1652, 2000.)

Free propagation speckle was registered for several values of the radial position r of the illuminated area. A number of 1024 signals sampled at 1 KHz were registered in each measurement. Although the GD (unweighted) measured at different radial positions did not show any noticeable dependence on the linear velocity of the illuminated region, the WGD with the nonzero weights in the closest 5 neighbors produced the results shown in Figure 5.8. A monotonic, close to linear dependence of it with the radius is observed.

5.5 FUJII'S METHOD

Fujii et al. have proposed a method for the register and processing of activity images using a scanning experimental set up.[9–10] It was originally intended for blood flow measurement.

Fujii uses a CCD linear sensor and a He–Ne laser expanded by a cylindrical lens to a line. The line, projected on the sample, is imaged onto the linear sensor.

When the speckle image intensity is successively scanned, the registered values change from one scan to the next as in consecutive frames in the aforementioned CCD images acquisition. As blood flow velocity increases, so does variation of intensity of the speckle pattern. The difference between output data for each pixel is then measured and integrated. It is repeated for more than a hundred of scans.

The spatial variation of blood flow level, or microcirculation map, may be visualized by plotting the integrated values.

The actual calculation is performed using Equation 5.3

$$D(n) = \sum_{k=1}^{N} \left| I_k(n) - I_{k+1}(n) \right| / [I_k(n) + I_{k+1}(n)] \tag{5.3}$$

where $I_k(n)$ is the pulse height of the output signal for the nth pixel point of the kth scan of the CCD sensor given by Equation 5.4.

$$I_k(n) = \int_0^{\Delta T} I(t) \, dt \tag{5.4}$$

where ΔT is the scanning interval in the sensor. This operation is named the *average difference, D(n).*

This measure is somewhat similar to the so-called *structure function* usually used for scattering measurements, but the square appearing in the structure function is replaced by the absolute value, and the result is weighted with the mean value of the two consecutive intensities.

The effect of this algorithm is to act as a bandpass filter on the incoming signal. The bandpass of the filter can be adjusted by modifying the integration time ΔT.

By adding a scanning system in the direction perpendicular to the illumination line, an operation generated by tilting a mirror, a 2-D map of the microcirculation can be obtained.

The image shown in Figure 5.9 was obtained using a CCD in place of the scanning system proposed by Fujii. Notice the high quality of the results. The full resolution

FIGURE 5.9 Application of Fujii´s algorithm. Notice the reduction of speckle and the high quality of the image.

of the CCD camera is achieved as no speckle can be perceived and little details, such as small balls in the rim, can be distinguished.

By using this method, the effect of the tuberculin test was used to depict an example of the performance of the method after 6 and 24 hours of application.[9]

An effect that is introduced by the denominator in Fujii´s algorithm is that different reflectivities, or illumination nonuniformities are treated in different ways, so that the same difference in the numerator results are amplified if the denominator is smaller. This is of no effect if the sample is uniformly illuminated and shows uniform reflectance but introduces artifacts if it is not. Notice how the dark regions corresponding to a rim in the shadows around the coin appear bright, indicating false high activity.

This boosting of activity effect in the darker regions should be taken into account for adequate interpretation of the results. Fujii patented this device.[11]

5.6 LASCA METHOD

In medicine, where the patient's body shows motion due to respiration, heartbeat, and other involuntary functions, there is considerable interest in developing techniques that can be implemented in real time.

Briers[12] has proposed and demonstrated a very important method, also intended for in vivo imaging blood perfusion in real time but that could be used for other phenomena exhibiting similar dynamics. It measures the spatial contrast in a time-integrated image.

In physiology,[13–14] perfusion is the passage of a fluid (in this case arterial blood) through the vessels of a certain organ through a capillary system in the tissue. Perfusion ("F") can be measured with the following formula, Equation 5.5

$$F = \frac{P_A - P_V}{R} \tag{5.5}$$

where P_A is mean arterial pressure, P_V is mean venous pressure, and R is called *vascular resistance*. Alternative methods for the measuring of perfusion require the use of functional magnetic resonance imaging (fMRI).

One of fMRI techniques used to measure tissue perfusion in vivo is based on the use of an injected contrast agent that changes the magnetic susceptibility of blood and thereby the MR signal, which is repeatedly measured during passage of medicine. The other relies on *arterial spin labeling* (ASL), where arterial blood water is tagged using radiofrequency pulses that alter the magnetization of the blood before it enters into the tissue of interest. The amount of labeling is measured and compared to a control recording obtained without spin labeling. Both alternatives are expensive and the former is invasive.

The measurement of perfusion is required in, to name a few, the treatment of diabetes, arteriosclerosis, vasospastic conditions associated with Raynaud's syndrome, burns, scars, wound healing, and other disorders.

There has been another approach to measure perfusion using a real time scanning[6] device. The measured magnitude, also called *perfusion*, is defined as the product of velocity and concentration of blood cells in the illuminated volume. It uses the power spectral density of the intensity as a measure of perfusion. Although this and some related techniques were developed for blood flow monitoring, some of them may also be useful for other applications.

The LASCA principle is based on an idea originating in the method developed by Elliasson and Mottier and on a previous similar analogical method by Briers requiring a photographic register and analogue processing.

5.7 THE ELLIASSON–MOTTIER METHOD

In 1971, Elliasson and Mottier[15] proposed a method to visually observe the nodal lines of a vibrating object illuminated by a speckle field. It was based on the blurring that occurs in the speckles in the image of an object during the integration time of the detector (the eye or camera) and is due to fast intensity changes. They registered a photographic record of the vibrating object as illuminated by a speckle field coming from a laser-illuminated diffuser.

The contrast of the obtained time-averaged recorded speckle patterns is high in the regions corresponding to nodes and decreases with vibration amplitude to zero. This method could be applied also to objects showing surface activity. The concept was improved by Fercher et al.[16] for flow visualization in the mapping of retinal blood flow. A photographic register of the time-averaged image speckle pattern was, after development, high-pass spatial filtered in an optical set-up to depict regions of equal blur.

Incoherent superposition of speckle patterns produces blur so that contrast in the resulting images diminishes. This is a concept also used in Synthetic Aperture Radar (SAR) images, which involve the addition of several speckle images of the same scene, named *looks*, to diminish their speckled appearance.

LASCA (laser contrast speckle analysis) uses the spatial first order statistics of time integrated speckle. This is an almost real time and nonscanning technique. If intensity variations are relatively fast, finite integration time in the acquisition system causes the standard deviation $\sigma_{x,y}$ of the (spatial) measured intensity I variations to diminish and so does the contrast defined by Equation 5.6.

$$C = \frac{\sigma_{x,y}}{\langle I \rangle} \tag{5.6}$$

where $\langle I \rangle$ is the spatial average of the intensity. This magnitude is a measure of the degree of blur due to the finite time of integration of the detector.

The functional dependence between C and the physical parameters related to the activity is not known except for a few particular cases.

As C diminishes, with increased activity and blur, the image constructed on this basis is a reversed contrast, conversely showing dark regions in active places. As a spatial standard deviation is required, the operation involves some reduction in resolution.

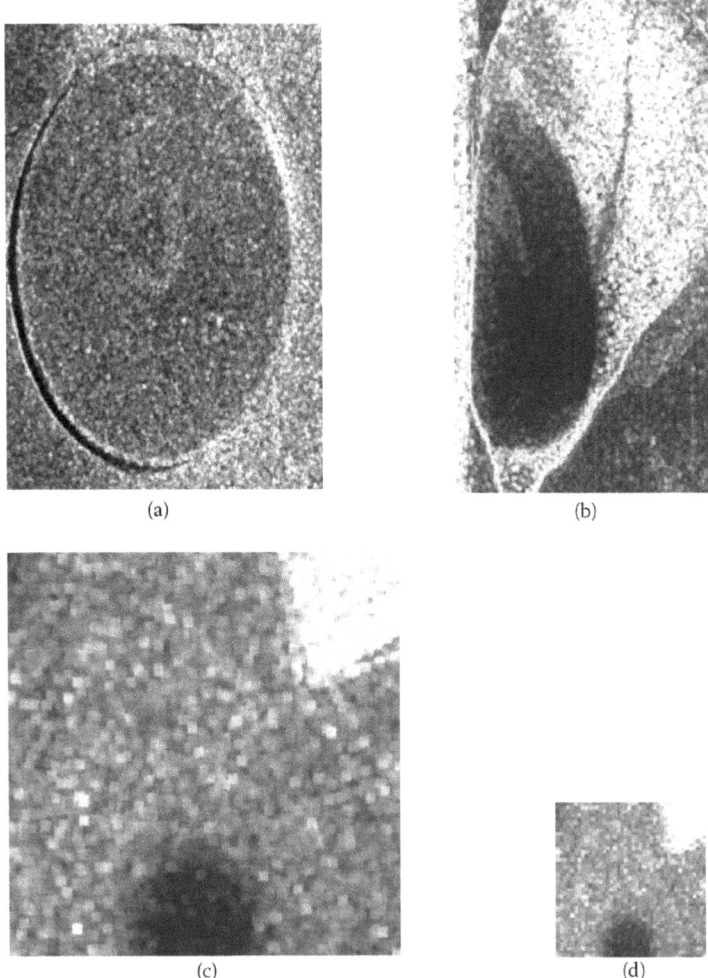

(a)   (b)

(c)   (d)

FIGURE 5.10 Results of applying the LASCA algorithm: (a) the paint on coin drying experiment, (b) a corn seed in germination process, (c) an apple bruised by letting it fall on steel ball in the middle lower side, and (d) a reduced resolution version of (c).

In Figure 5.10a the result of the coin covered with paint is shown. As it can be seen LASCA method is not adequate for this type of experiment. To compare with other images of the coin notice that the contrast in LASCA images is reversed. Dark regions then indicate high activity. The topography of the coin under the paint almost cannot be perceived and resolution is low. Notice the artifact consisting in a double bright line surrounding a dark one in the right side due to the use of a variance window in the sharp discontinuities produced by the shadow.

Figure 5.10b shows a corn seed reproducing the high activity area in dark grey, and low activity area in light grey. In this case, LASCA algorithm was able to distinguish the alive tissue, in dark, from the food storage portion, in light.

Figures 5.10c and 5.10d show the results obtained when LASCA is applied to a bruising of fruit experiment[17] (expanded and original resolution are shown, respectively). In this experiment the contrast was the tool that could determine the bruised region for a longer period of time. The light grey in the upper right corner of the images is related to an inert piece of metal included as a nonactive reference.

Space variance of intensity can be related to the time integral of the time auto-covariance of the intensity for ergodic phenomena (here, this term refers to phenomena where space statistical measures are also representative of time statistic measures).

The spatial variance of the intensity σ^2 is found by using Equation 5.7.

$$\sigma_s^2 = \frac{1}{T} \int_0^T C_V(\tau) \, d\tau \tag{5.7}$$

where

$$C_V(\tau) = \langle (I(T) - \langle I \rangle_t)(I(t+\tau) - \langle I \rangle_t) \rangle \tag{5.8}$$

is the time autocovariance of the intensity. Assuming ergodicity,[18] $C_V(\tau)$ reduces to Equation 5.9.

$$C_V(\tau) = \langle I \rangle^2 C_t(\tau) \tag{5.9}$$

being $C_V(\tau)$ the time autocorrelation function of the intensity.

The higher the velocity of the scatterers (or other high activity reasons), the faster the autocorrelation of the intensity falls and thus the lower the space contrast is in the time integrated image.

$C_V(\tau)$ is determined by the physical origins of the dynamic speckle and is related to the broadened power spectrum of the scattered light by the Wiener Kintchine theorem through a Fourier transform.

The Wiener–Kintchine theorem states that the power spectral density of a wide-sense-stationary random process is the Fourier transform of the corresponding autocorrelation function.

So that, if by experiment or analysis the auto-correlation can be assumed to be known or measured, then the behavior of LASCA images can be predicted and related to physical origin or previous experimental knowledge.

When the scattered light shows a Lorentzian power spectrum, then the autocorrelation is a pure negative exponential as a consequence of the Wiener–Kintchine theorem. The contrast can then be analytically calculated by Equation 5.10.[16]

$$k = \frac{\sigma_s}{\langle I \rangle} = \left\{ \frac{\tau_c}{2T} \left[1 - \exp\left(\frac{-2T}{\tau_c}\right) \right] \right\}^{1/2} \tag{5.10}$$

where T is the exposure time of the CCD and τ_c is the correlation time that, for blood perfusion measurements, is given by Equation 5.11.

$$\tau_c = \frac{1}{(ak_0 v)} \qquad (5.11)$$

where a is a factor that depends on the Lorentzian width and scattering properties of the tissue,[19] k_0 is the wave number, and v the velocity of the scatterers.

The left side of Equation 5.11 is the LASCA measurement. A finite window of $N \times N$ pixels size is selected in the time-integrated, registered image of the dynamic speckle, and the contrast is calculated from the intensity values I inside the window. The result is assigned to the central pixel of the window in a new image. Then, the window is displaced N pixels, and the procedure is repeated for the entire image.

τ_c can then be deduced from these measurements involving only first order statistics. Figure 5.11 shows an example of cerebral blood flow obtained using the LASCA technique.

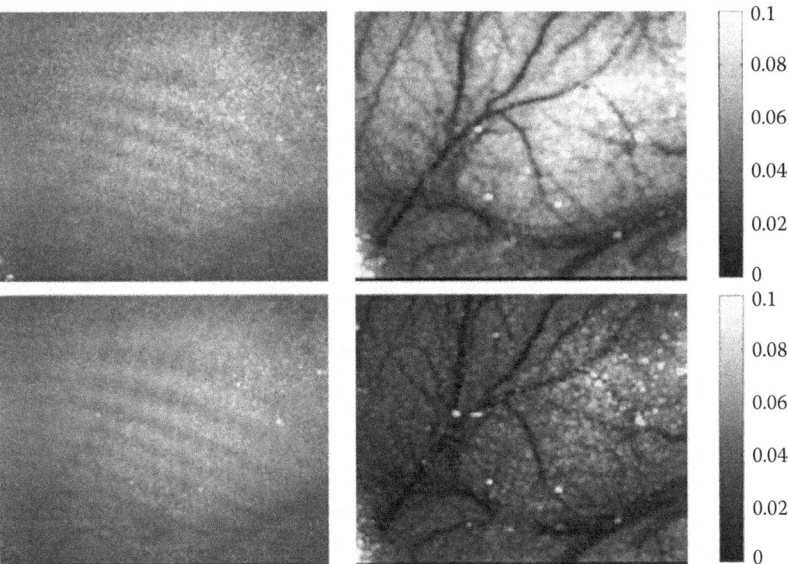

FIGURE 5.11 Raw speckle images (left column) and the corresponding speckle contrast images (right column) computed directly from the raw speckle images with 5 × 5 areas of pixels. The top row shows typical images under normal blood flow and the bottom row corresponds to increased blood flow conditions. Low speckle contrast values (dark areas in right column) correspond to increased speckle blurring of the raw speckle images (left column) indicating increased blood flow. Darker areas in the speckle contrast in bottom right image indicate increased cerebral blood flow in the microvasculature compared with normal conditions (top right). Each image shows a 5 × 4 mm area of cortical surface. (Reprinted by permission from Macmillan Publishers Ltd: *Journal of Cerebral Blood Flow and Metabolism* 21, Andrew K. Dunn, Hayrunnisa Bolay, Michael A. Moskowitz, and David A. Boas. Dynamic imaging of cerebral blood flow using laser speckle, 195–201, ©2001.)

For erythrocytes, the following approximate expression has been proposed[20] to obtain their velocity v_c, given by Equation 5.12.

$$v_c \approx \frac{3.5}{\tau_c}\mu ms^{-1}$$
(5.12)

When the analytic expression relating contrast to physical origins of the activity is known, such as Equation 5.12 for blood cells, the optimal exposure integration time can be deduced. As the method depends heavily in the choice of the integration time of the detection, it has been studied for blood flow changes by Shuai Yan et al.[21]

Although this measure has been designed specifically for some physical phenomena, namely blood perfusion, it can be expected that all phenomena giving rise to a certain activity will cooperatively diminish the contrast. So, it gives a qualitative estimation of the time scales involved. The use of several different integration times could help to distinguish different phenomena that in this version could be pooled together.

For example, the LASCA measure has also been used to measure depth-resolved multiple scattering[22] in phantoms aiming to the visualization of microcirculation in human tissues on a point-by-point measurement basis, using fiber optics both for illumination and detection.

An approach similar to LASCA was proposed by Konishi el al.[23] It was implemented in real time to visualize retinal microcirculation. They found that an adequate measure for flow-graphy was a different measure BR defined as in Equation 5.13. Instead of the standard deviation, the average deviation is used, and the reciprocal of a measure similar to the contrast is calculated. It is also called the SNR, blur rate, or reciprocal of speckle contrast.

$$BR_{k,j} = \frac{\langle I_{k,j}\rangle}{(1/N)\sum_{n=1}^{N}\left|I_{k,j}(t_n) - \langle I_{k,j}\rangle\right|}$$
(5.13)

where

$i_k(t, s)$ is the intensity detected in the k-th pixel of the detector with area σ; $I_k(t_n)$ is defined in Equation 5.14 and its average value by Equation 5.15.

$$I_k(t_n) = \int_{\sigma} ds \int_{t_n}^{t_n + \Delta T} i_k(t,s)dt$$
(5.14)

$$\langle I_k\rangle = \frac{1}{N}\sum_{n=1}^{N} I_k(t_n)$$
(5.15)

Inverse contrast images are obtained and the measure results to be almost proportional to the blood flow velocity. Figure 5.12 shows a result obtained with the same

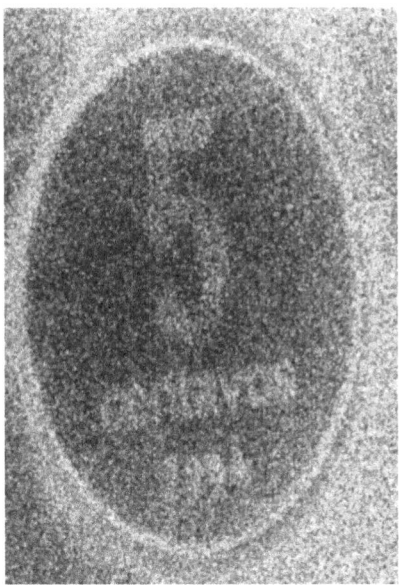

FIGURE 5.12 Result obtained with the Konishi algorithm.

coin used before, where several details of the topography can be noticed but the image is very speckly.

5.8   OTHER IMAGE METHODS

5.8.1   TIME CONTRAST

Contrast of the intensity values along time has also been used to build activity images. Temporal contrast is defined as:

$$C = \frac{\sigma_t}{\langle I \rangle_t} \tag{5.16}$$

where the brackets indicate temporal mean value of the intensities in each pixel and σ_t the corresponding standard deviation.

This contrast measurement requires several images, so that the acquisition system should be able to resolve the intensity variations, but does not show reduced spatial resolution. Its use was demonstrated by imaging obscured subsurface inhomogeneities[24] and also was applied to measure blood flow through the skull of a live rat.[2]

Figure 5.13 shows temporal contrast images of the painted coin and of a bruised apple. Notice that the topographic details of the coin can scarcely be perceived in this image. In the bruised apple, the bruising also is not very visible but a clear dot of unknown origin that was not visible in the apple appears slightly upwards. This dot also appears in other types of processing.[25] There are still detection possibilities that have not been fully explored.

(a)

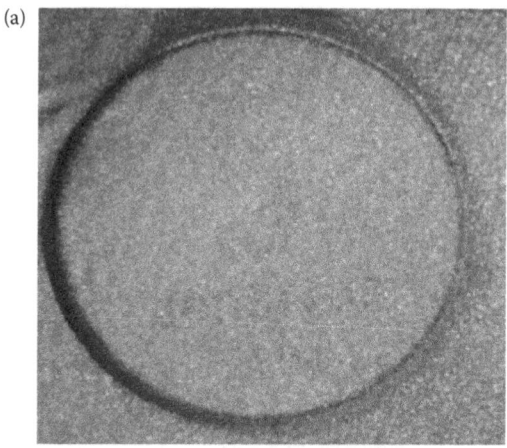

(b)

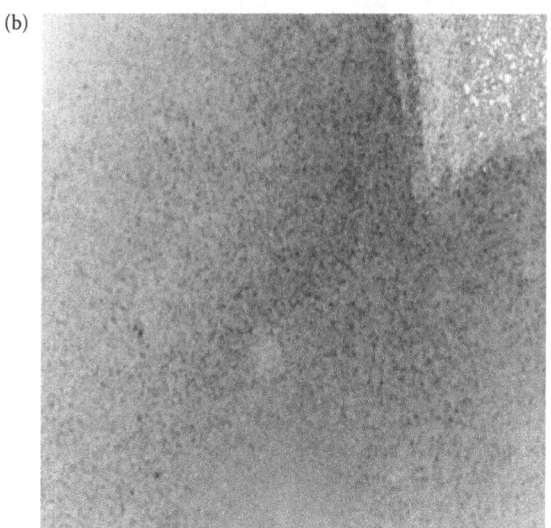

FIGURE 5.13 Temporal contrast images (a) of the painted coin and (b) of the bruised apple.

5.9 HETERODYNE APPROACH OF ATLAN ET AL.

A very interesting and somewhat different approach was recently proposed by
Atlan et al.[26–27] They propose the assessment of cerebral blood flow (CBF) changes
with a wide-field laser Doppler imager based on a CCD camera detection scheme, in
vivo, in mice. A lensless Fourier Transform digital holography set-up with acousto-
optic modulators in the reference beam is used to introduce heterodyne detection.
The Doppler signature of moving scatterers is measured in the frequency domain.
The quadratic mean of the measured frequency shift is used as an indicator of CBF.
Very good images were obtained that allowed the observation of a significant variability of that indicator in an experiment designed to induce blood flow changes.

Other recent activity image display methods have been reported in the literature that admit some physical interpretation. These are the Empirical Mode Decomposition[25] and the variance of phase one.[28]

5.10 CONCLUSIONS

We have reviewed some techniques to display the loci of equal activity in images of dynamic speckle. They were selected in a rather arbitrary way. These techniques do not display the same aspects of the phenomena and cannot, in general, be replaced one by the other. Rather, in some aspects, some of them show some desirable features, such as the possibility of real or almost real-time implementation, easiness of interpretation, and the possibility to assign some physical meaning or shortcomings: reduced resolution, high number of required images, long processing time, high storage requirements, artifacts such as boosting dark regions activity, etc.

So, some techniques are better suited for an application than others, as can be seen in the examples shown here.

For many cases (seeds, fruits, etc.) the inner dynamics of the phenomena are unknown or poorly known, so some experimentation is required to choose the best suited technique for every new application.

There are other approaches, involving time resolved images and showing different aspects of the time evolution of the signals in each pixel, that are going to be treated in the forthcoming chapters.

REFERENCES

1. Dunn, A., Devor, A., Bolay, H., Andermann, M., Moskowicz, M., Dale, A. and Boas, D., Simultaneous imaging of total cerebral hemoglobin concentration, oxygenation, and blood flow during functional activation, *Opt. Lett.*, 28(1), 28, 2003.
2. Li, P., Ni, S., Zhang, L., Zheng, S. and Luo, Q., Imaging cerebral blood flow through the intact rat skull with temporal laser speckle imaging, *Opt. Lett.*, 31(12), 1824, 2006.
3. Romero, G., Alanís, E. and Rabal, H., Statistics of the dynamic speckle produced by a rotating diffuser and its application to the assessment of paint drying, *Opt. Eng.*, 39(6), 1652, 2000.
4. Cheng, H., Luo, Q., Wang, Z., Gong, H., Chen, S., Liang, W. and Zeng, S., Efficient characterization of regional mesenteric blood flow by use of laser speckle imaging, *Appl. Opt.*, 42(28), 5759, 2003.
5. Passoni, I., Modelos en Bioingeniería: Caracterización de imágenes estáticas y dinámicas, Tesis Doctoral, Universidad Nacional de Mar del Plata, Mar del Plata, 2005.
6. Wardell, K., Jakobson, A., and Nilsson, G., Laser perfusion imaging by dynamic light scattering, *IEEE Trans. Biomed. Eng.*, 40(4), 309, 1993.
7. Arizaga, R., Cap, N., Rabal, H. and Trivi, M., Display of the local activity using dynamical speckle patterns, *Opt. Eng.*, 41, 287, 2002.
8. Rabal, H., Ortiz, C., Trivi, M., and Arizaga, R., Activity speckle measurements comparison using R.O.C. (receiver operating characteristic) methods, *in Proc. of SPIE*, SPIE, 2006, 6341 63411B-1.

9. Fujii, H., Nohira, K., Yamamoto, Y., Ikawa, H. and Ohura, T., Evaluation of blood flood by laser speckle image sensing Part 1, *Appl. Opt.*, 25(24), 5321, 1987.
10. Fujii, H., Asakura, T., Nohira, K., Shintomi, Y. and Ohura, T., Blood flow observed by time-varying laser speckle, *Opt. Lett.,* 10(3), 104, 1985.
11. Fujii, H., Apparatus for monitoring bloodstream, Patent No. CA1293535, 1991.
12. Briers, J. D. and S. Webster, Laser speckle contrast analysis (LASCA): A nonscanning full field technique for monitoring capillary blood flow, *J. Biomed. Opt.*, 1, 174, 1996.
13. Briers, J. D., Time varying speckles for measuring motion and flow, *Saratov Fall Meeting 2000: Coherent Optics of Ordered and Random Media,* Zimnyakov, D. A., Ed., *Proc. SPIE*, 2000, 4242, 25.
14. Cardiovascular Physiology Concepts, Richard E. Klabunde, Ph.D., http://www.cvphysiology.com/Hemodynamics/H001.htm.
15. Eliasson, B. and Mottier, F., Determination of the granular radiant distribution of a diffuser and its use for vibration analysis, *J. Opt. Soc. Am.,* 61(5), 559, 1971.
16. Fercher, A. and Briers, J., Flow visualization by means of single exposure speckle photography, *Opt. Comm.*, 37(5), 326, 1981.
17. Pajuelo, M., Baldwin G., Arizaga, R., Cap, N., Rabal, H., Trivi, M., Bio-speckle assessment of bruising in fruits, *Opt. Laser Eng.*, 40, 13, 2003.
18. Goodman, J. W., Some effects of target-induced scintillation on optical radar performance, *Proc. of IEEE*, 53(11), 1688, 1965.
19. Bonner, R. and Nossal, R., Model for laser Doppler measurements of blood flow in tissue, *Appl. Opt.*, 20, 2097, 1981.
20. Richards, G. and Briers, J. D., Capillary blood flood monitoring using speckle contrast analysis (LASCA): Improving the dynamic range, *Proc. SPIE*, Tuchin, V. V., Podbielska, H., and Ovryn, B., Eds., 1997, 160.
21. Yuan, S., Devor, A., Boas, D., Dunn, A., Determination of optimal exposure time for imaging of blood flow changes with laser speckle contrast imaging, *Appl. Opt.*, 44(10), 1823, 2005.
22. Gonik, M. M., Mishin, A. B., and Zimnyakov, D. A., Visualization of blood microcirculation parameters in human tissues by time-integrated dynamic speckles analysis, *Ann. New York Acad. Sc.*, 972, 325, 2002.
23. Konishi, N. and Fujii, H., Real time visualization of retinal microcirculation by laser flowgraphy, *Opt. Eng.*, 34(3), 753, 1995.
24. Nothdurft, R. and Yao, G., Subsurface imaging obscured subsurface inhomogeneity using laser speckle, *Opt. Expr.*, 13(25), 10034, 2005.
25. Federico, A. and Kaufmann, G. H., Evaluation of dynamic speckle activity using the empirical mode decomposition method, Alejandro Federico, Guillermo H. Kaufmann, *Opt. Comm.*, 267(2), 287, 2006.
26. Atlan, M., Gross, M., Forget, B., Vitalis, T., Rancillac, A., and Dunn, A., Frequency-domain wide-field laser Doppler in vivo imaging, *Opt. Lett.*, 31(18), 2762, 2006.
27. Atlan, M., Forget, B. C., and Boccara, A. C., Cortical blood flow assessment with frequency-domain laser Doppler microscopy, *J. B. Opt.*, 12(2), 2007.
28. Rabal, H., Cap, N., Trivi, M., Arizaga, R., Federico, A., Galizzi, G. and Kaufmann, G., Speckle activity images based in the spatial variance of the phase, *Appl. Opt.*, 45, 8733, 2006.

6 Frequency Analysis

Lucía Isabel Passoni, Gonzalo Hernán Sendra, and Constancio Miguel Arizmendi

CONTENTS

6.1 FREQUENCY SIGNAL ANALYSIS OF SPECKLE PATTERNS

The aim of this section is to provide a comprehensive glance of different methods used to characterize the time series evolution of the dynamic speckle patterns within the paradigm of frequency and time-frequency analysis.

Several authors have accomplished different processes based on the frequency analysis of these time series, finding encouraging results.

6.2 THE FOURIER TRANSFORM

Frequency analysis is, roughly speaking, the process of decomposing a signal into frequency components, that is, complex exponential signals or sinusoidal signals. If we adopt a deterministic signal model for analyzing time-varying speckle signals, the mathematical tool for spectral analysis is the Fourier transform. The Fourier transform of a signal $x(t)$ is expressed as

$$X(\omega) = \int_{-\infty}^{\infty} x(t)e^{-j\omega t} dt \qquad (6.1)$$

where

$$\omega = 2\pi f$$

$|X(\omega)|$ amplitude of each ω component

$\angle\{X(\omega)\}$: phase shift of each ω component

$X(\omega)$ exists if $x(t)$ satisfies the *Dirichlet conditions*, requiring that $x(t)$: have a finite number of maxima or minima within any finite interval, have a finite number of discontinuities within any finite interval, and be absolutely integrable, that is,

$$\int_{-\infty}^{\infty} |x(t)| \, dt < \infty \qquad (6.2)$$

The signal $x(t)$ can be synthesized from its *spectrum* $X(\omega)$ by using the following inverse Fourier transform equation

$$x(t) = \frac{1}{2\pi} \int_{-\infty}^{\infty} X(f)e^{j2\pi ft} df \qquad (6.3)$$

The energy of $x(t)$ can be computed in either the time or frequency domain using the Parseval's relation

$$E_x = \int_{-\infty}^{\infty} |x(t)|^2 \, dt = \int_{-\infty}^{\infty} |X(\omega)|^2 \, d\omega \qquad (6.4)$$

The function $|X(\omega)|^2 \geq 0$ shows the distribution of energy of $x(t)$ as a function of frequency. Hence, it is called the *energy spectrum* of $x(t)$.

6.2.1 FOURIER TRANSFORM OF DIGITAL SIGNALS

In most practical applications, discrete-time signals are obtained by sampling continuous-time signals periodically in time. It can be mathematically expressed as a convolution between the signal and a periodic impulse train

$$x[n] = x(t) \sum_{n=0}^{N-1} \delta(t - nT) = \sum_{n=0}^{N-1} x(nT)\delta(t - nT) \tag{6.5}$$

where T is the sampling period, and δ is the Dirac (Kronecker) delta function. The quantity $f_s = 1/T$, the number of samples taken per unit of time, is called the sampling rate or sampling frequency.

The Fourier transform of expression results in

$$X(\omega) = \sum_{n=0}^{N-1} x[n]e^{-j\omega n} \tag{6.6}$$

Then, the synthesis equation becomes

$$x[n] = \frac{1}{2\pi} \int_{-\pi}^{\pi} X(\omega)e^{j\omega n} d\omega \tag{6.7}$$

6.2.2 DISCRETE FOURIER TRANSFORM (DFT)

The N-points discrete Fourier transform (DFT) of an N-point sequence $x[n]$ is defined by

$$X[k] = \sum_{n=0}^{N-1} x[n]e^{-j(2\pi/N)kn} \quad k = 0, 1, \ldots, N-1 \tag{6.8}$$

The sequence $x[n]$ can be recovered from its DFT coefficients $X[k]$ by the following DFT formula:

$$x[n] = \frac{1}{N} \sum_{n=0}^{N-1} X[k]e^{j(2\pi/N)kn} \quad n = 0, 1, \ldots, N-1 \tag{6.9}$$

The Discrete Fourier transform (DFT) of finite duration digital signal is equal to the Fourier transform of the signal at frequencies

$$\omega_k = \frac{2\pi}{N} k \qquad 0 \le k \le N-1 \tag{6.10}$$

There are faster algorithms for the calculation of the DFT, known as the Fast Fourier transform (FFT).

The use of the DFT requires some considerations.[1-2]

- The only way to improve the frequency resolution is using more signal samples in the DFT algorithm:

- The sampling frequency f_s must be equal to at least twice the signal band-width to avoid *aliasing effect* due to high frequencies superposition pro-duced by the convolution implicit in the sampling process.[1,3]
- Band-limited signals do not exist in practice. Therefore, it is necessary to include a low-pass frequency filter to diminish high frequency content over $f_s/2$ as much as possible. In the case of light intensity sampling with a CCD camera (as other light detectors), the exposure time acts as a low-pass filter.[3]
- The $x[n]$ sequence is zero outside the interval of observation. This is equivalent to multiplying the signal with a rectangular window. Hence, the resulting spectrum is the convolution between the signal and window spectrums. Apart from that, as the signal is considered to be periodic by the DFT algorithm, if there are differences between its first and last sample values, the spectrum will have an increment in the high frequencies. The use of alternative window shapes can reduce this effect.[1]

6.3 FILTERING THE SPECTRUM

We mentioned before that the spectrum contains useful information distributed in frequency. Sometimes we would be interested in filtering or equalizing this spectrum to attain better results or to find a particular frequency band. When the speckle signal is generated by a mixture of different speckle dynamics, such as biospeckle, some appropriate frequency filtering or frequency-dependent amplification may be necessary.[4]

There are several types of filters depending on the characteristic of the spectral (magnitude and phase) response. Some elementary continuous (analog) low-pass filters are briefly introduced as follows.[5]

- *Butterworth filter.* Butterworth low-pass filter tries to approximate an ideal square filter with the equation:

$$H(\omega) = \frac{1}{1 + \omega^{2N}} \qquad (6.11)$$

where N is the filter order.

It is characterized by maximally flat magnitude response over the pass-band.
- *Chebyshev filter.* Chebyshev filter tries to approximate the ideal low-pass response with the equation:

$$H(\omega) = \frac{1}{1 + \varepsilon^2 T_N^2(\omega)} \qquad (6.12)$$

where $T_N(\omega)$ is a Chebyshev polynomial of order N.

Chebyshev filters present a steeper roll-off than Butterworth filters, although they also contain ripple in the pass band. The inverse Chebyshev filter is a variation in which the pass band is maximally flat, with a ripple in the stop-band. Both Chebyshev approximations permit high selective pass band filter implementations with fewer amounts of calculations than the equivalent Butterworth one.

- *Elliptic filter.* Elliptic approximation (also known as the Cauer approximation) is the most commonly used function in the design of filters. The elliptic approximation is a rational function with finite poles and zeros, whereas the Butterworth and Chebyshev approximations have all their zeros at infinite. In particular, in this function the location of the poles must be chosen to provide an equiripple characteristic in both the stop and pass bands. Thus, for a given requirement, the elliptic approximation will generally require a lower order than Butterworth or Chebyshev filters.

Any low-pass filter approximation can be transformed into a high-pass, a band-pass, or a stop-band filter one, easily modifying the cut-off frequencies.[5]

Digital filters can be categorized into two broad classes known as finite impulse response (FIR) and infinite impulse response (IIR) filters. The IIR filter is similar to an analog filter in which the response to a unit impulse decays exponentially over time but never disappears completely. The usual approach to the design of IIR filters involves the transformation of an analog filter into a digital filter (bilinear transformation) meeting the prescribed specifications.

There are many reasons for designing FIR filters. It is easy to design a FIR filter that approximates a prescribed magnitude but retains a linear phase (it preserves the shape of the original waveform). In addition, the nonrecursively characteristic of FIR filters assures its stability. The round-off noise associated with FIR realizations can easily be made small. However, the major disadvantage of FIR filters is that a high length impulse response is required to approximate sharp cutoff filters, compared to an equivalent IIR filter. Two common methods for designing FIR filters are window-based filter design and the equiripple approximation.[6,1]

6.4 POWER DENSITY SPECTRUM

The second-order moments of a stationary random sequence, the autocorrelation function and the power density spectrum (PSD), play a crucial role in signal analysis. In this section, we introduce the estimation of the PSD using a finite data record of the process. The power density spectrum is defined as the Fourier transform of the autocorrelation function r with k lags, $r(k)$:

$$S(\omega) = \sum_{k=-\infty}^{\infty} r(k)e^{-i\omega k} \tag{6.13}$$

Another definition is

$$S(\omega) = \lim_{N\to\infty} E\left\{ \frac{1}{N} \left| \sum_{n=0}^{N-1} x[n]e^{-j\omega n} \right|^2 \right\} \tag{6.14}$$

with $E\{\}$ representing the mathematical expectation and N the signal sample length.

6.4.1 PSD ESTIMATION

The PSD is defined using continuous signals. When we are dealing with finite-length digital signals, it is necessary to estimate the PSD. There are many estimators, such as the periodogram or correlogram estimation, Blackman–Tukey, Barlett–Welch, Daniell, Thompson multitaper, etc. A brief description[2] is

- *Periodogram estimation.* Neglecting the expectation and the limit in eq. (6.14), we get

$$\hat{S}_P(\omega) = \frac{1}{N}\left|\sum_{n=0}^{N-1} x[n]e^{-j\omega n}\right|^2 \tag{6.15}$$

- *Correlogram estimation.* It is based on the autocorrelation definition

$$\hat{S}_C(\omega) = \sum_{k=-(N-1)}^{N-1} \hat{r}[k]e^{-i\omega k} \tag{6.16}$$

- *Blackman–Tukey estimation.* Both the periodogram and the correlogram present the disadvantage of high leakage and smearing of the spectrum,[6] resulting in unacceptable variance (asymptotically unbiased estimator). To reduce this effect, a locally smoothed PSD is obtained by weighting the estimated autocorrelation with a lag window (or taper) function $w[k]$.

$$\hat{S}_{BT}(\omega) = \sum_{k=-(M-1)}^{M-1} w[k]\hat{r}[k]e^{-i\omega k} \tag{6.17}$$

There is a tradeoff between resolution and variance. The variance reduction depends on the windows choice,[2]
- *Barlett–Welch estimation.* The basic idea of the Barlett method is to reduce the large fluctuation of the periodogram by splitting up the available record of N observations into $L = N/M$ data segments of M observation each, and then average all corresponding periodograms.

$$x_i[n] = x[(i-1)M + n - 1] \quad \begin{array}{l} n = 0,\ldots,M-1 \\ i = 1,\ldots,L \end{array}$$

$$\hat{S}_B(\omega) = \frac{1}{L}\sum_{i=1}^{L}\left\{\frac{1}{M}\left|\sum_{n=0}^{M-1} x_i[n]e^{-j\omega n}\right|^2\right\} \tag{6.18}$$

The Welch estimation differs from the Barlett estimation because it allows overlapping data segments. Although the resolution is limited by the segment length, the variance diminishes as the number of segments increases. This is one of the most frequently used PSD estimation methods. Figure 6.1 shows the PSD Welch estimator of two THSP series

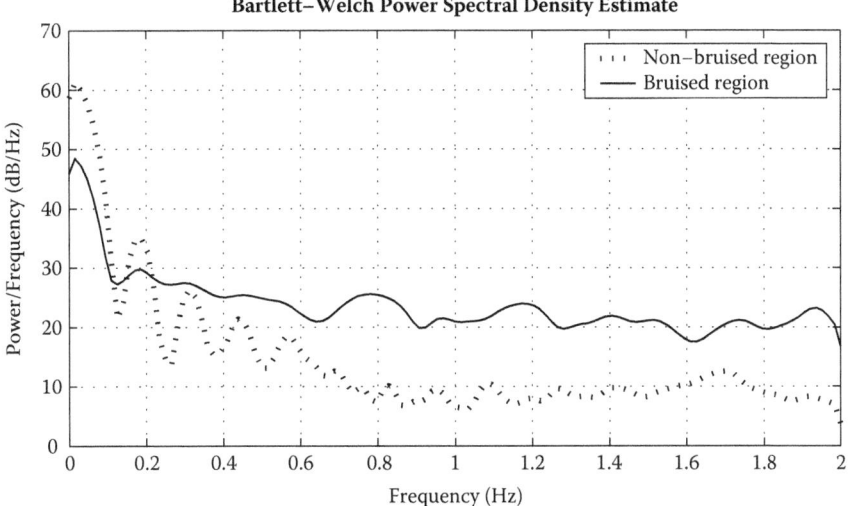

FIGURE 6.1 PSD of a THSP corresponding to an apple-bruised region.

$$x_i[n] = x[(i-1)K + n - 1] \quad \begin{array}{l} n = 0, \ldots, M-1 \\ i = 1, \ldots, S \end{array}$$

$$\hat{S}_W(\omega) = \frac{1}{S} \sum_{i=1}^{S} \left\{ \frac{1}{MP} \left| \sum_{n=0}^{M-1} v[n] x_i[n] e^{-j\omega n} \right|^2 \right\} \tag{6.19}$$

$$P = \frac{1}{M} \sum_{n=0}^{M} |v[n]|^2$$

- *Thompson multitaper estimation.* In this method, several windows (tapers) are used on the same data record to compute several modified periodograms. These modified periodograms are then averaged to produce the multitaper spectral estimate. The central premise of this multitaper approach is that if the data tapers are properly designed orthogonal functions, then, under mild conditions, the spectral estimates would be independent of each other at every frequency. Thus, averaging would reduce the variance, while proper design of full-length windows would reduce bias and loss of resolution. Thompson suggested windows based on discrete prolate spheroidal sequences (also known as *Slepian* tapers) that forms an orthonormal set, although any other orthogonal set with desirable properties can also be used.[2]

6.5 TIME-FREQUENCY ANALYSIS

In numerous applications the signal analysis using the Fourier transform is extremely useful, because its frequency components allow the signal characterization. However, in the transformation process, the time information is totally lost. When observing the Fourier transformed signal, it is impossible to determine when a certain event takes place. If the signal is stationary, it implies that its properties don't change in time, so this limitation is not too important. However, most of the signals contain nonstationary or transitory characteristics: tendencies, abrupt changes, beginnings and finalizations of events that are often the most interesting aspects to be analyzed. In these cases, the Fourier analysis is not the most appropriate detection method. Such fact immediately points out the need for a time-frequency representation of a signal that would give us local time-frequency information. In the Fourier case, it is obvious that we need a more local waveform to achieve this. The most intuitive way to overcome such obstacle is to localize the sinusoids in the Fourier presentation by windowing its application scope, using a window with compact support.

Time-localization can be achieved by first windowing the signal so as to cut off only a well-localized slice and then taking its Fourier transform. This gives rise to the short-time Fourier transform (STFT) or windowed Fourier transform. The magnitude of the STFT is called the *spectrogram*.

6.5.1 THE SHORT-TIME FOURIER TRANSFORM

The short-time Fourier transform of a signal $x(t)$ using a window function $g(t)$ is defined as follows.

$$STFT(f,s) = \int_{-\infty}^{\infty} x(t)g(t-s)e^{-j2\pi ft}dt \tag{6.20}$$

Considering the window $g(t)$ as sliding along the signal $x(t)$ and for each shift $g(t-s)$, the usual Fourier transform of the product function $x(t)\, g(t-s)$ is computed.

With the purpose of treating this problem, Gabor adapted the Fourier transform to analyze short sections of the signal per time, thus obtaining the windowed Fourier transform or the short-time Fourier transform (STFT). It maps the signal into a function of two dimensions: time and frequency.[7]

The STFT maintains a commitment relationship among the time and frequency projections. It offers information on both time and frequencies of the event happening. Nevertheless, it is not possible to obtain arbitrarily fine localization in time and in frequency due to the uncertainty principle. Note that the precision is determined by the size of the window. Although this relationship allows the localization of events of interest, the problem is that, once chosen, the window size stays fixed for all the frequencies. The signal analysis often requires a more flexible focus to more precisely determine time and frequency of occurrence of events of interest.

With STFT we have fixed both levels of time and frequency localizations. As a next step in the time frequency analysis, Wavelet Transform is presented.

6.5.2 WAVELET TRANSFORM

In the late 1970s, the work of Jean Morlet, a geophysical engineer at the French oil company Elf Acquitaine, came up with an alternative for the STFT. His development of a tool for oil prospecting is usually taken as the starting point of the history of wavelets.[8]

The standard procedure, which consists of first windowing a signal and then computing its Fourier coefficients, amounts to taking the inner product of the signal with windowed Fourier functions depending on two parameters—the location of the window and the frequency label of the different coefficients. The signals that Morlet wanted to analyze did consist of different features in time and frequency, which he wanted to disentangle.[9]

A small window, in the windowed STFT, is "blind" to low frequencies, which correspond to signals too large for the window. On the other hand, large windows lose information about a brief change. To overcome this trouble Morlet kept constant the number of oscillations and either stretched or compressed the window.

Morlet brought the idea to generate the transform functions in a different way: to get a higher frequency analysis wavelet he compressed a cosine wave windowed with a Gaussian function, and to obtain a lower frequency function he stretched it out, this wavelets could be also shifted in time. Hence, this transform function depends on two different parameters: location and compression scale. The signal analysis is performed using the inner product of the signal with the shifted and compressed (or expanded) wavelet. In consequence wavelets allow finer time localization in highest frequencies. This process allows fixing long intervals of time when one needs more precise information on low frequencies, and shorter regions when information of high frequencies is required. Figure 6.2 shows the contrast among the different analysis: time, frequency, STFT, and Wavelets.

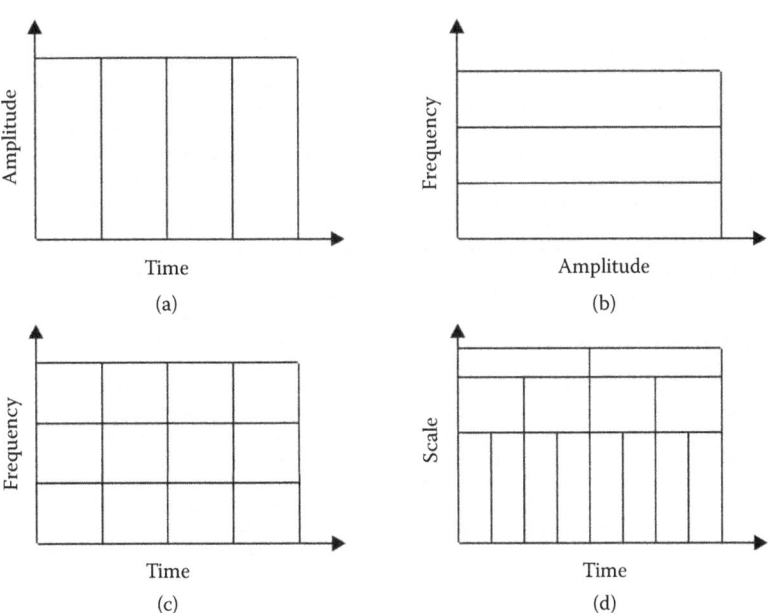

FIGURE 6.2 Signal analysis domain: (a) time, (b) frequency (Fourier analysis), (c) time and frequency (STFT), and (d) time and scale (wavelet analysis).

Wavelet transform analysis allows revealing data features that other analysis techniques lose, such as tendencies and discontinuities. It facilitates a different vision from the data of those presented by traditional techniques. This analysis can be used for compression or noise reduction without appreciable degradation.

One of the essential advantages of Wavelet transforms is the ability to carry out local analysis of a particular region in an extensive signal. Starting from the enunciated advantages, it is necessary to consider the foundation of this analysis. The generic characterization of a "wavelet" refers to a limited-duration signal with zero mean. Comparing waves with sinusoid signals that are the base of the analysis of Fourier, it is inferred that sinusoids have no limited duration (they extend from $-\infty$ to $+\infty$). Although sinusoids are soft and predictable, wavelets spread to become irregular and asymmetric.

In a similar way to the Fourier analysis that separates the original signal in sinusoids of different frequencies, the analysis with wavelets divides the signal in shifted and scales versions of the mother wavelet. Signals with abrupt changes could be better analyzed with an irregular wavelet than with a soft sinusoid. It still makes more sense to think that local features are better characterized by compact support wavelets.

The process of Fourier analysis represented by Fourier is defined as the integral of signal $f(t)$ multiplied by an exponential complex function.

$$X(\omega) = \int_{-\infty}^{\infty} x(t)e^{-j\omega t}dt \tag{6.21}$$

The Fourier transforms are the coefficients $X(\omega)$ that multiplied by the sinusoid of ω frequency give the constituent waves of the original signal. The wavelet analysis is a method sustained by the characterization of the signal by its amplitude distribution in an appropriate base. If it is required that the base be an orthogonal one, any arbitrary function can be decomposed in a unique form and its decomposition can be inverted.[10] The wavelet analysis is an appropriate tool to detect and characterize specific phenomena in the planes of frequency and time.

The wavelet is a smooth function with an oscillation that disappears quickly with good localization either in frequency or in time. A family of wavelets $\psi_{a,b}$ is the group of elementary functions generated by the shifting and delays of only a mother wavelet.

$$\psi_{a,b}(t) = |a|^{-1/2}\psi\left(\frac{t-b}{a}\right) \tag{6.22}$$

where $a,b\in\Re$ $a\neq 0$ are, respectively, the scale and localization parameters and t is the time. In consequence, it is possible to obtain their replicas in different scales and locations, thus starting an initial pattern.

The continuous wavelet transform $(W\psi)$ of a signal $x(t)$ is defined as the function correlation among $x(t)$ with the family of wavelets $\psi_{a,b}$ for each a and b.

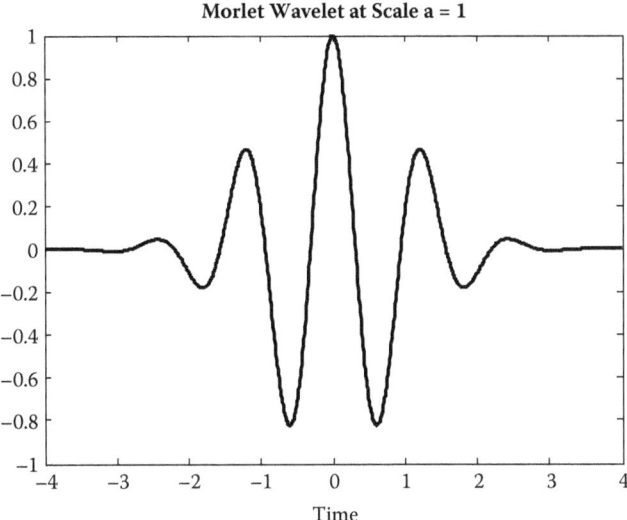

FIGURE 6.3 Morlet wavelet at scale $a = 1$.

$$W_\psi x(a,b) = |a|^{-1/2} \int_{-\infty}^{\infty} x(t)\psi^*\left(\frac{t-b}{a}\right) dt = \langle x, \psi_{a,b} \rangle \qquad (6.23)$$

Where ψ^* is the complex conjugate of ψ, a is the frequency dilation factor and b the time translation parameter.

The results of the continuous wavelet transform are the coefficients function of the scale and position. Multiplying each coefficient with the wavelet appropriately scaled and displaced, the constituent waves of the original signal are obtained.

Different wavelets are proposed as the Haar (a one period square wave) or the Morlet wavelet expressed as:

$$\psi(t) = e^{-\frac{t^2}{2}} \cos(5t) \qquad (6.24)$$

It is also a small wave because the Gaussian exponential is close to zero outside the interval $-3 < t < 3$, presented in Figure 6.3

The graph of $\psi((t-b)/a)$ is obtained stretching the graph of $\psi(t)$ by the factor a, called the scale, and shifting in time by b. Figure 6.4 shows a signal $x(t)$ along with the Morlet wavelet at three scales and shifts.

We can assume CWT as the inner product or cross correlation of the signal $x(t)$ with the scaled and time-shifted wavelet. This cross correlation is a measure of the similarity between the signal and the scaled and shifted wavelet. This point of view is illustrated in Figure 6.4.

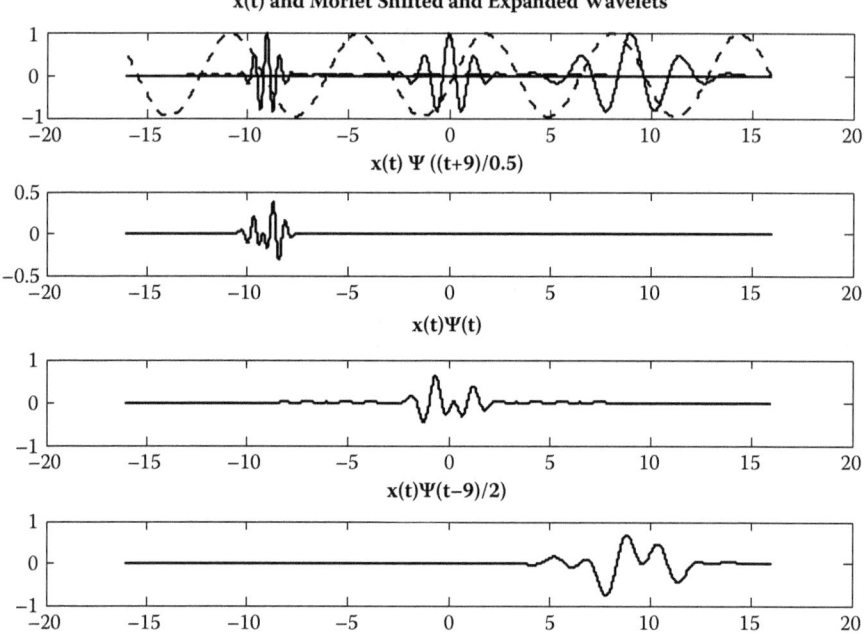

FIGURE 6.4 Product between the time signal and the shifted and expanded Morlet wavelets.

Continuous wavelet transform (CWT) provides a time-scale description similar to the STFT with a few important differences[11]:

- Frequency is related to the scale that may have a better relationship with the problem.
- CWT is able to resolve both time and scale (frequency) events better than STFT.

In some cases we can get an orthogonal basis of functions in CWT case by choosing the scales to be powers of 2 and the times to be an integer multiple of the scales.

Selecting the mother wavelet and a discrete group of parameters $a_j = 2^j$ and $b_j = 2^j k$ with $j, k \in Z$ (integer set), the family

$$\psi_{j,k}(t) = 2^{-j/2}\psi(2^{-j}t - k), \quad j,k \in Z \qquad (6.25)$$

is an orthonormal basis in the Hilbert space $L^2(\Re)$ of finite energy signals.

Calculating wavelet coefficients at every possible scale generates a lot of unnecessary data. If a subset of scales and positions is chosen to make calculations, a compact wavelet analysis can be performed with similar results. Scales and positions based on powers of two (dyadic scales and positions) are selected to turn the analysis efficient and just as accurate.

With the restriction of a discrete set of parameters we get the discrete Wavelet transform (DWT), which corresponds to an orthogonal basis of functions, all derived from a single function called the *mother wavelet*. The basis functions in DWT are not solutions of differential equations as in the Fourier case. The basis functions are "near optimal" for a wide class of problems. This means that the analysis coefficients drop off rapidly. There is a connection and equivalence to filter bank theory from digital signal processing that leads to a computationally efficient algorithm. Such an analysis is obtained from the discrete wavelet transform (DWT). Discrete wavelet transform (DWT) generates a nonredundant representation of the signal, and the values of $\langle X, \psi_{a,b} \rangle$ are the coefficients. These coefficients offer information organized in a hierarchical outline of nested subspaces called multiresolution analysis, making possible the estimate of energy in different scales.

It is supposed that the original signal comes from sampled values being carried out its decomposition over all the resolution levels $N = log_2(M)$, the expansion results:

$$X(t) = \sum_{j=0}^{\infty} \sum_{k} C_j(k)\psi_{j,k}(t) = \sum_{j=0}^{\infty} r_j(t) \qquad (6.26)$$

where the coefficients $C_j(k)$ can be interpreted as the local residual errors between successive signal approximations at scales j and $j + 1$, and $r_j(t)$ is the residuals signal at scale j, containing information of the original sign corresponding to the frequencies $2^{j-1}\omega_s \leq |\omega| \leq 2^j \omega_s$.

An efficient way to implement this scheme using filters was developed in 1988 by Mallat. The Mallat algorithm is in fact a classical scheme known in the signal processing community as a *two-channel subband coder*.

The filtering process, at its most basic level, looks like the scheme presented in Figure 6.5. The original signal, X, passes through two complementary filters and emerges as two signals: residuals and approximation. If this operation is performed on a real digital signal, we will obtain twice the original data. Hence, downsampling is performed, keeping only one point out of two in the decomposition path.

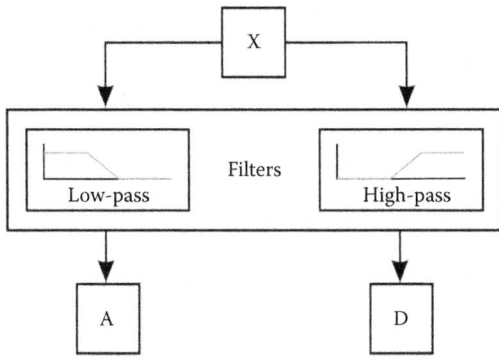

FIGURE 6.5 Wavelet filtering process.

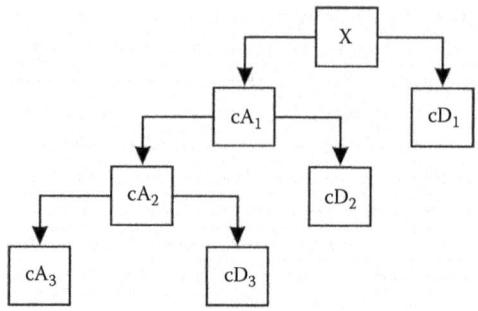

FIGURE 6.6 Wavelet decomposition tree.

The decomposition process can be iterated with successive approximations being decomposed in turn, so that one signal is broken down into many lower resolution components. This is called the *wavelet decomposition tree*. See Figure 6.6.

Because the analysis process is iterative, in theory it can be continued indefinitely. In reality, the decomposition can proceed only until the individual details consist of a single sample. Figure 6.7 shows a THSP ($X(t)$) decomposition performed with the Daubechies wavelet in five levels, obtaining five details series and an approximation one. The original $s(t)$ series can be reconstructed as

$$X = a_5 + d_5 + d_4 + d_3 + d_2 + d_1 \tag{6.27}$$

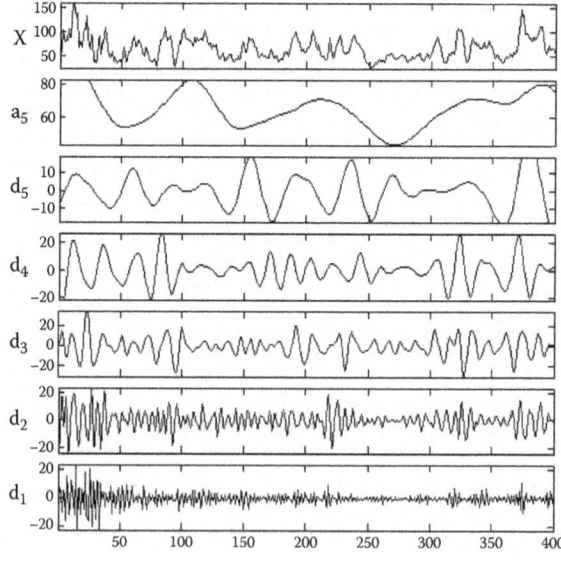

FIGURE 6.7 Multiscale representation of wavelet transform of $X(t)$, in 5 levels, using Daubechies wavelet ($n = 7$) $X = a5 + d5 + d4 + d3 + d2 + d1$.

6.6 DESCRIPTORS BASED ON FILTERED SIGNALS

6.6.1 ENERGY

One of the simplest descriptors of a signal is its energy. For a finite length digital signal, it can be approximated by

$$E = \sum_{n=0}^{N-1} x^2[n] \tag{6.28}$$

When this descriptor is applied to filtered speckle signals, information regarding phenomena is often present at different frequency bands. The energy corresponding to each band could then be used to determine some properties of the sample.[12] The same procedure can also be used with the power density spectrum, where the spectral band power would be used as a descriptor.

The energy of some particular spectral bands proved to be useful in the detection of bruising damage in fruits,[12] the study of corn seeds, where vitreous and floury regions of the endosperm were able to be discriminated by the energy of some spectral bands,[13] and the identification of chemotactic responses in swarm assays.[14] In these cases, fifth-order Butterworth filters were employed to filter each pixel speckle signal. The energy of each band was used to make a pseudocolored image. Figure 6.8 shows three images obtained from corn kernel corresponding to the following spectral bands: (a) 0 to 0.1Hz, (b) 0.2 to 0.3Hz, and (c) 0.8 to 0.9Hz.[14]

6.6.2 HIGH TO LOW FREQUENCY RATIO (HLR)

This descriptor was first proposed by Fujii to estimate the skin blood flow,[15] and is given by

$$\text{HLR} = \frac{|V_H|^2}{|V_L|^2} \tag{6.29}$$

where $|V_H|^2$ and $|V_L|^2$ indicate the powers of certain high and low frequency components of speckle signals obtained through different band-pass filters. This parameter was able to reflect quite sensitively the change of blood flow in the physiological test called *reactive hyperaemia* (response of blood flow change from reduced flow to reopened flow).[16]

6.6.3 DESCRIPTORS BASED ON THE PSD

6.6.3.1 Mean Frequency

For analyzing different spectral shape, Aizu and Asakura proposed the use of the mean frequency $\langle f \rangle$ defined by

$$\langle f \rangle = \frac{\sum_i f_i P(f_i)}{\sum_i P(f_i)} \tag{6.30}$$

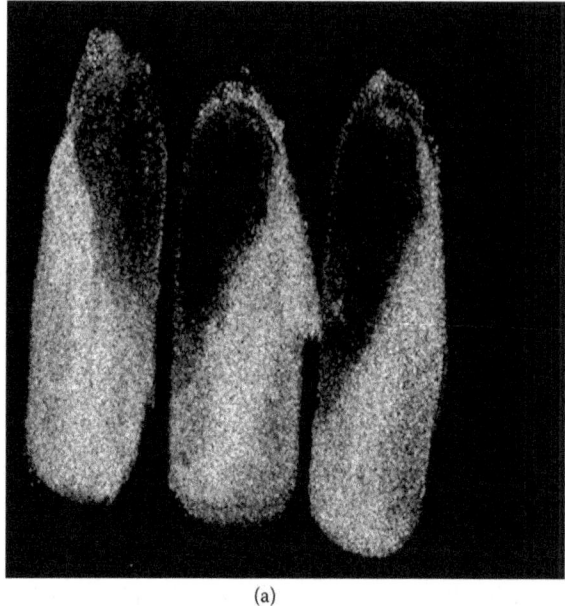

(a)

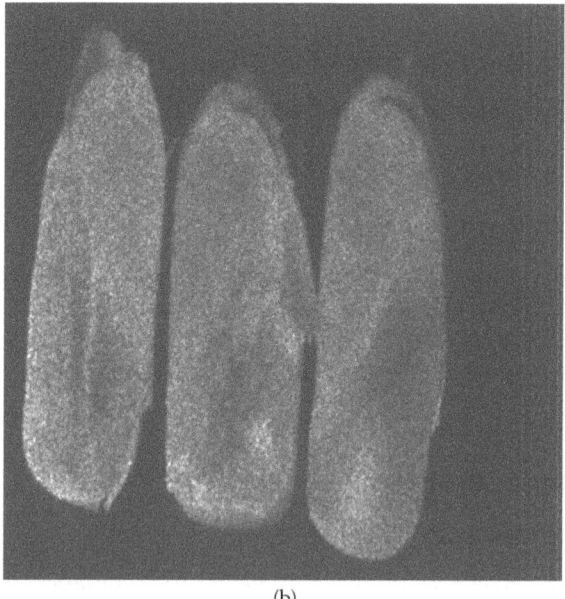

(b)

FIGURE 6.8 Corn kernel pseudocolored energy images of three spectral bands: (a) 0 to 0.1 Hz, (b) 0.2 to 0.3 Hz, and (c) 0.8 to 0.9 Hz.

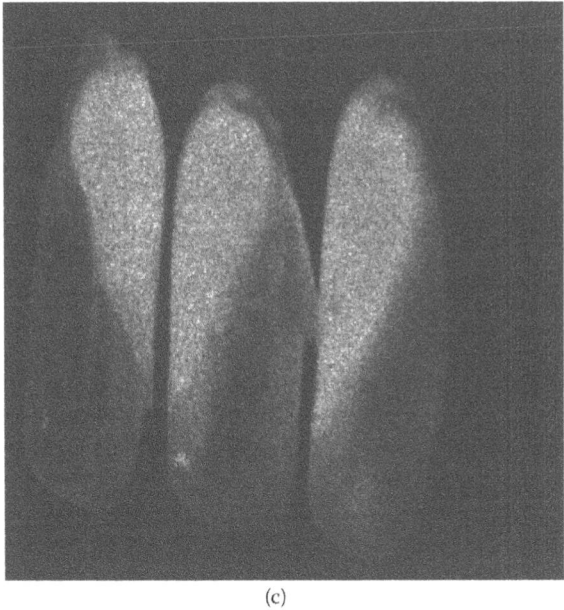

(c)

FIGURE 6.8 (*Continued*)

where $P(f_i)$ is the signal power at frequency f_i.[16] The experiments show that this descriptor successfully measured blood flow variation with vasoconstrictor and vasodilator.

6.6.3.2 Modified Mean Frequency

Ruth estimated the blood flow using what he called a "blood flow parameter"[17] that can be interpreted as modified mean frequency parameter:

$$M = \sqrt{\int T^2(f)P(f)df}$$ (6.31)

where $P(f)$ is the signal power spectrum and $T(f)$ is the transfer function or weighting function. Strictly speaking, the blood flow parameter is not proportional to the average blood flow, but can be optimized by choosing an appropriate transfer function to better reflect the blood flow changes. The choice of the best transfer function is made empirically because it depends on the type of skin tissues and involuntary body movements.[18]

6.6.3.3 Cut-Off Frequency

The cut-off frequency descriptor is defined as the frequency giving half the value of the power spectrum at zero frequency. Power spectral distributions of speckle signals under different conditions of blood flow in gastric mucous membrane were obtained by Aizu and Asakura.[16] The cut-off frequency was able to discriminate between these conditions.

Aizu and Asakura measured the cut-off frequency from the power spectra of three different inanimate objects in motion with normal and strong diffuseness.[4] The latter produced small speckle grains because light was scattered not only from its surface but also from its inside scatterers (multiple scattering). The effect finally causes higher frequency fluctuations in the detected signals. In conclusion, power spectra are dependent on the scattering structure of objects in which multiple scattering occurs.

6.6.3.4   Descriptors Based on the Wavelet Transform

6.6.3.4.1   Wavelet Entropy-Based Descriptor

Claude Shannon introduced the information-theoretic entropy in 1948.[19] According to the information theory, entropy is a relevant measure of order and disorder in a dynamical system. By using entropy, no specific distribution needs to be assumed. The spectral entropy, as defined by the Fourier power spectrum, shows a natural approach to quantify the degree of order of a complex signal, indicating the spread level of the signal power spectrum.[20] A regular activity, like a tone, shows a peak in the frequency domain. This concentration of the frequency spectrum in one single peak corresponds to a low entropy value. On the other extreme, a nonregular activity noise will show components in a wide band range of the frequency domain, thus being reflected in higher entropies. The stationary condition to apply the Fourier transform (FT) is not ensured in the THSP. To deal with these limitations, time evolving entropy can be defined from a time-frequency representation of the signal, as provided by the wavelet transform WT.[21]

The discrete wavelet transform (DWT) makes no assumptions about signal stationary feature. Therefore, although entropy based on the WT reflects the degree of order/disorder of the signal, it can provide additional information about the underlying dynamical process associated with the signal.[22]

Wavelet transform provides a tool to observe a time series at a full range of different scales a, while retaining the time dimension of the original data. Multiresolution analysis theory shows that no information is lost if the continuous wavelets coefficients are sampled at a sparse set of points in the scale-time plane known as the *dyadic grid*. This grid leads to the discrete wavelet transform (DWT) where the scale parameter is $a_j = 2^{-j}$ and the translation $b_{j,k} = 2^{-j}k$, with $j, k \in Z$.

In order to study the biospeckle temporal evolution, the THSP rows are divided into temporal windows of length L and for each segment i ($i = 1, \ldots N_T$; with $N_T =$ *signal length/L*).

The following expression is used to obtain the mean wavelet energy of the detail signal j at each time window i

$$E_j^{(i)} = \frac{1}{N_j} \sum_{k=0}^{(L/2^j)-1} |C_{k,j,i}|^2 \text{ with } i = 1, \ldots, N_T \tag{6.32}$$

where $C_j(k)$ can be interpreted as the local residual errors between successive signal approximations at scales j and $j + 1$, and $r_j(t)$ is the residual signal at scale j, containing signal information corresponding to the scales $2^{j-1}\omega_s \leq |\omega| \leq 2^j\omega_s$, and where N_j represents the number of wavelet coefficients at resolution level j included in the time interval i.

The total energy at interval i can be obtained by:

$$E_{total}^{(i)} = \sum_{j<0} E_j^{(i)} \tag{6.33}$$

The THSP row window i relative wavelet energy will be given by:

$$p_j^{(i)} = \frac{E_j^{(i)}}{E_{total}^{(i)}} \tag{6.34}$$

The expression given by Blanco[23] is used to evaluate the Shannon entropy in window i. The obtained value is assigned to the central window point.

$$S_{WT}^{(i)} = -\sum_{j<0} p_j^{(i)} \cdot \ln\left[p_j^{(i)}\right] \tag{6.35}$$

Consequently, it is possible to characterize the signal by the time evolving entropy value. This parameter can be considered as a descriptor of the dynamic biospeckle.

The wavelet entropy S_{WT} of the speckle time series is proposed to characterize the biospeckle phenomenon of a THSP image and also of a set of frames where the dynamic biospeckle shows no-uniform activity regions.

The former method was applied to a series of experiments.[23] Namely, to the characterization of slow and fast paint drying process and to evaluate activity images applied to viability test of corn seeds, as well to detect nonvisible bruising in apples.

6.7 PAINT-DRYING EXPERIMENTS

The dynamic speckle patterns corresponding to different drying stages in samples of synthetic paint were registered after extending them on a glass surface by means of an extender. The samples were illuminated with an expanded and attenuated 10 mW He–Ne laser. The speckle images were then registered by a CCD camera, digitized to 8 bits by a frame grabber and stored in the memory of a personal computer. In order to register the data, the technique described by Oulamara[24] was used. Two experiments were performed, one employing paint with slow drying and the other with fast-drying paint.

In the first case, for every state of the phenomenon being assessed, 512 successive images of the dynamical speckle pattern are registered, and a certain column is selected in each of them. By selecting the image middle column, a THSP is constructed. In the case of the slow-drying paint, the THSP is assumed to be a representative sample of the state of the drying phenomenon being assessed when it was registered.

A set of 23 THSP of 512 × 512 pixels images was obtained to evaluate the drying of a latex painted sample. The sampling frequency was 12.5Hz; consequently, the acquisition time of a whole THSP lapses 40 s. The time interval between two consecutive drying stages is about 4 min. It is assumed that a particular drying stage can be characterized by the time series analysis of the THSP image rows. Consequently, the intensity amplitude from each image row is converted into a time series to be processed using the wavelet entropy (WE) descriptor approach.

Since drying of latex does not suffer any significant change during the time of acquisition of each image THSP, WE was calculated using a window with the whole row width. A single descriptor of THSP image was obtained with the average of the WE corresponding to the 512 lines.

For testing the descriptor performance, the WE value of each THSP was plotted versus the drying time.[23] The decay rate of the WE plot looks very similar to the evolution of the sample weight loss (gravimetrical measurement), a conventional method used in paint technology. Weight loss was measured with an analytical balance (±0.1 mg). Figure 6.9 shows the WE values normalized to a 0–1 scale, the gravimetrical measurements, and also the second-order moment (inertia moment) calculated by Amalvy et al. Hence, it is shown that the WE points follow the measurement

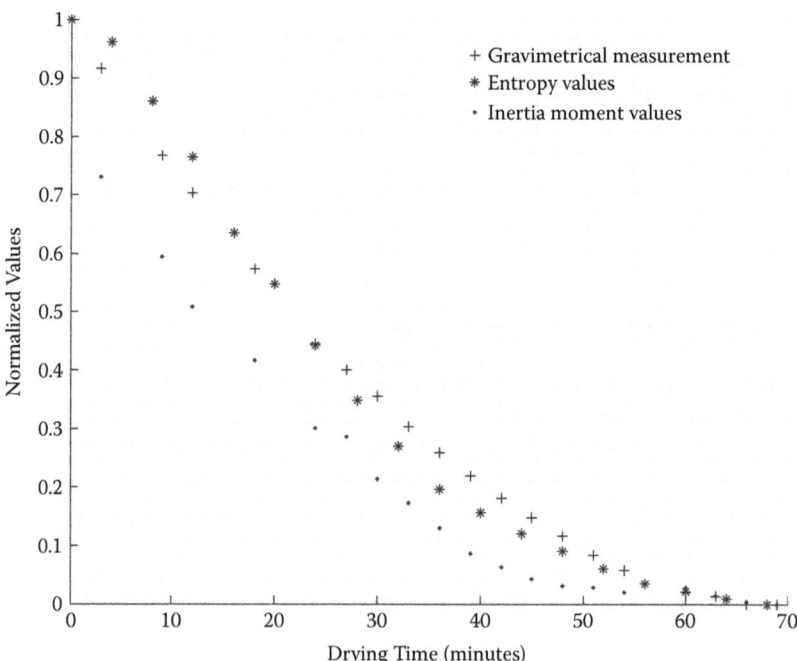

FIGURE 6.9 THSP entropy mean value, gravimetrical measurements, and inertia moment (second-order moment of the co-occurrence matrix). (Reprinted from *Optics Communications*, 246, Passoni, I. et al., Dynamic speckle processing using wavelets based entropy, 219–228, Copyright (2005), with permission from Elsevier.)

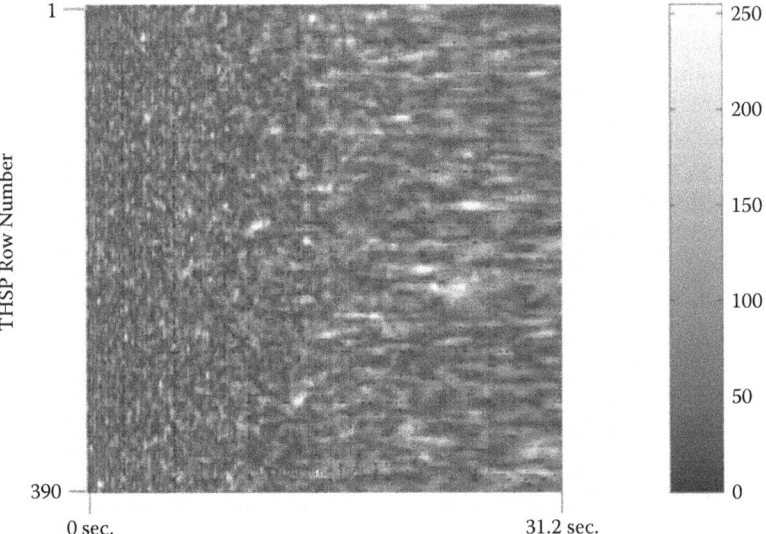

FIGURE 6.10 Nail enamel THSP obtained with 390 rows and 390 time samples. Sampling rate 0.08 s. Intensity level out of a 256 gray scale. (Reprinted from *Optics Communications*, 246, Passoni, I. et al., Dynamic speckle processing using wavelets based entropy, 219–228, Copyright (2005), with permission from Elsevier.)

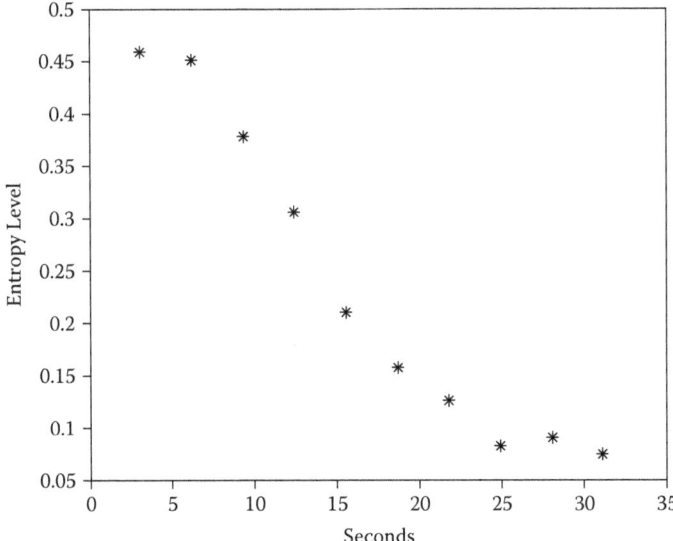

FIGURE 6.11 Nail drying process. (Reprinted from *Optics Communications*, 246, Passoni, I. et al., Dynamic speckle processing using wavelets based entropy, 219–228, Copyright (2005), with permission from Elsevier.)

values better than the inertia moment ones. According to the theories, the characteristic regions of the drying process show an initial stage of high and constant rate change, followed by a second zone where the rate of change is slower, and a third zone where the rate of change is a minimum.[25]

A relatively faster phenomenon, namely the drying of nail enamel was also tested. In this case, most of the drying process occurs during the register of a single THSP. Figure 6.10 shows this THSP. The whole drying process is shown in that image, as can be appreciated in the growing of speckle lifetime from left (wet state) to right (dry state). Consequently, it is a representation of a nonstationary dynamic speckle pattern. Then, the application of the entropy calculation must be performed in a slightly different way.

In this case, every THSP row is divided into nonoverlapped time windows. A three-level decomposition using the Daubechies (order 2) orthogonal wavelet is accomplished on each segment. An entropy mean value corresponding to each time segment was calculated, considering the whole row set. Figure 6.11 shows the plot of the WE values vs. drying elapsed time. The descendent entropy slope evidenced a time activity change during the first eight segments ($\approx$ 25 sec.); during this lapse the main drying process is performed. Afterwards, the entropy exhibits a rather constant behavior.

6.7.1 ACTIVITY IMAGES

When extended regions of a sample are investigated, usually their activity is not uniform. Then, it is sometimes necessary to segment the images according to their activity level. There are several image processing methods for this purpose, and they have proved to be useful in several applications. (See Chapter 5).

The following sections are devoted to evaluate activity images from biological dynamic speckle with wavelet entropy.

6.8 CORN SEED VIABILITY TEST

A sequential image set with 100 samples of speckle obtained from a whole corn seed (previously hydrated and bisected) is assembled. The size of each image exhibits a resolution of 480 × 250 pixels. Hence, a total number of 120,000 THSP (one corresponding to each pixel) was obtained.

The WE descriptor is calculated with a time windows of 100 samples, given that the stationary condition of the speckle pattern is assumed (acquisition time is shorter than the lapse where the condition of the viability seed could change). Hence, WE values are obtained corresponding to the time series of each image pixel, permitting the regions segmentation with different activity levels.

Figure 6.12 shows the result of processing the seed biospeckle; the different gray level regions are weighted according to the WE values.[23] It is clearly observed that high activity regions (brighter areas) correspond to higher descriptor (embryo region); meanwhile, medium image values (middle gray sites) indicates the presence of living cells, and the lower descriptor values (darkest regions) correspond to

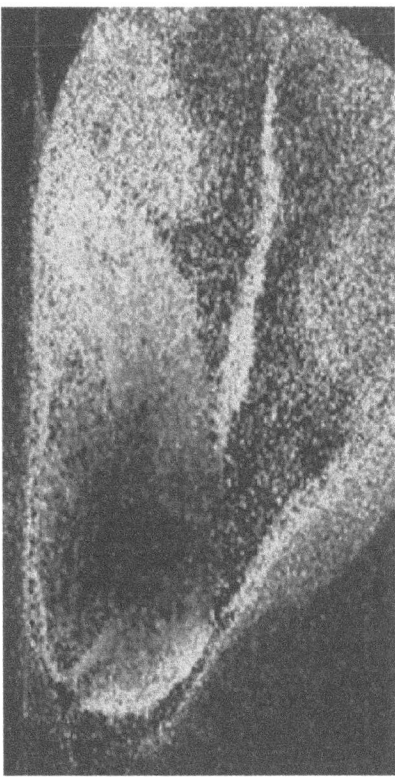

FIGURE 6.12 Corn seed activity image obtained using WE pixel values. (Reprinted from Optics Communications, 246, Passoni, I. et al., Dynamic speckle processing using wavelets based entropy, 219–228, Copyright (2005), with permission from Elsevier.)

the scene background. These data and their processing of a corn seed are discussed further in Chapter 8.

6.9 BRUISING IN APPLES

In order to detect bruised apple regions, a sequence of 500 whole field 300 × 300 speckle images were obtained from red delicious apples. The damage was caused by a controlled impact produced by letting fall a steel ball on the apple, and the extent of the damage could not be appreciated by visual inspection.

The image series is assembled into a 3-D array, hence 90,000 THSP are processed to attain a biospeckle descriptor array with pixel resolution under WE view. The process enables the segmenting of regions with different bioactivity levels, essentially bruised and nonbruised zones. Figure 6.13 shows the higher amplitude region that corresponds to an almost circular shape (middle bottom) centered in the bruised

Entropy Scale

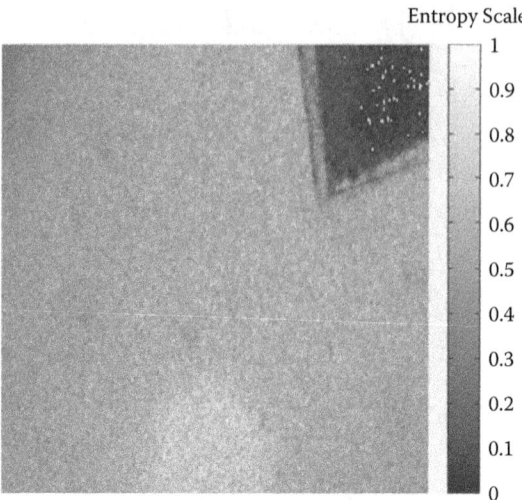

FIGURE 6.13 Bruised apple WE processed activity image. Bruising (brighter region, high entropy levels) at middle bottom and inert region at right top (lowest entropy values, darker zone). (Reprinted from Optics Communications, 246, Passoni, I. et al., Dynamic speckle processing using wavelets based entropy, 219–228, Copyright (2005), with permission from Elsevier.)

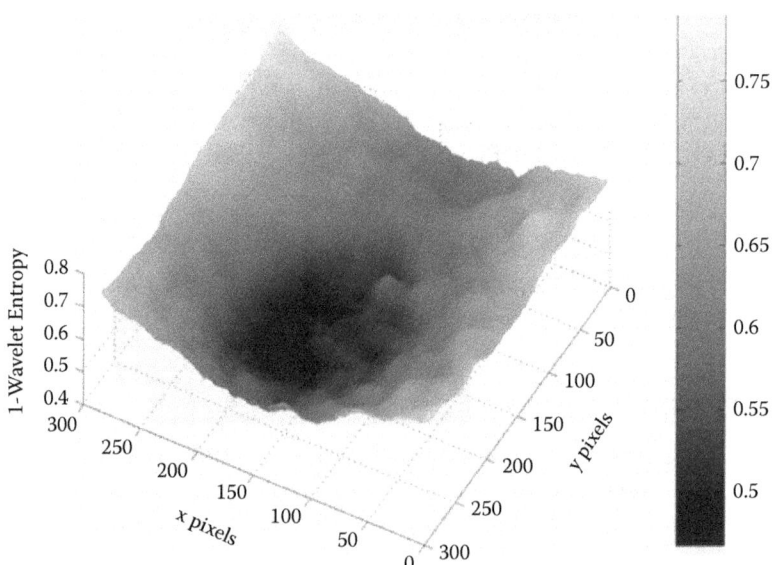

FIGURE 6.14 Bruised apple WE processed activity image without inert region. Bruised region (darkest, lowest (1-WE) values) looks like a depression. (Reprinted from *Optics Communications*, 246, Passoni, I. et al., Dynamic speckle processing using wavelets based entropy, 219–228, Copyright (2005), with permission from Elsevier.)

region, where an activity difference can be distinguished from the undamaged surrounding site. Also, at the upper right corner, an inert piece of material was inserted to test the biospeckle descriptor value, and as result, a very low (lower than the non-bruised apple zone) region is displayed.

Another experiment was carried out with a bruised apple, obtaining a set of 500 images of 300 × 300 pixels each. In this case, the image did not exhibit an inert zone (like the previous one). Hence, the whole 90,000 THSP were processed with the WE algorithm. In order to enhance the result visualization, the obtained WE image was smoothed using a 2-D median filter. Then, a complement function (1-WE) was applied and a 3-D view was plotted. Hence, the bruised area looked like a depressed zone (see Figure 6.14).

6.10 SEED DRYING

The study of viable and dead seeds losing water, with the use of wavelets, has been presented by Braga et al.[26] Sets of dead and viable bean seeds (*Phaseolus vulgaris L.*) were prepared by wetting them for 12 h. The seeds were illuminated on the inner part of a cotyledon every 4 h, for six illuminations. (See additional details in Chapter 8.)

The results for the seeds are presented in Figure 6.15, where, it is possible to compare the entropy from wavelets coefficients and the coefficients of inertia moment (IM). Both methods are adequate to separate dead from viable seeds. Wavelet transform has an extra advantage, the possibility of performing a time-frequency analysis of the water loss.

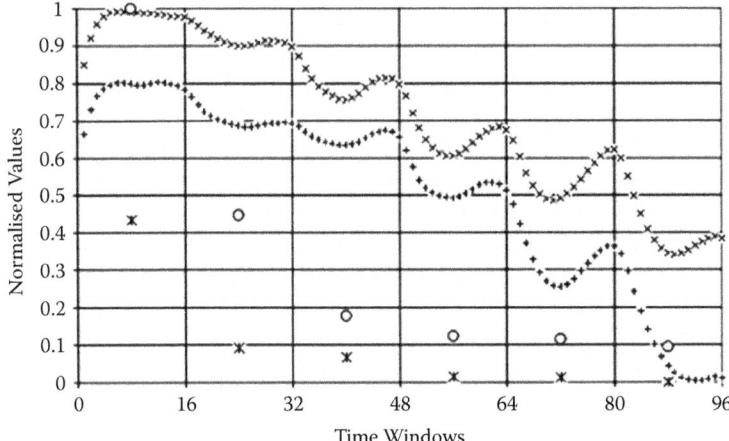

FIGURE 6.15 IM and entropy values for the seed drying process, standardized between 0 and 1, where "×" is entropy of alive seeds; "+" the entropy of dead seeds; "0" the IM of alive seeds; "*" is IM of dead seeds. (Reprinted from *Computers and Electronics in Agriculture*, 58, Braga, R.A. et al., Biological feature isolation by wavelets in biospeckle laser images, 123–132, Copyright (2007), with permission from Elsevier.)

6.11   THSP TEXTURE FEATURE VECTORS

The extraction of a set of texture feature vectors using the 2-D wavelet transform to characterize the time evolution of dynamic speckle patterns was proposed by Fernández Limia et al.[27]

Texture feature vectors from THSP patterns are calculated from the three details coefficients (horizontal, vertical, and diagonal) c_{ij} of the development of the THSP image in a selected wavelet basis. Feature vectors are obtained using modulus, energy, standard deviation, and the mean residual, calculated for every terminal frequency channel i, where M is the number of coefficients associated with the subband i and vm_i is the mean value of the M coefficients of the frequency channel i.

$$md_i = \frac{1}{M} \sum_{j=1}^{M} |c_{ij}| \tag{6.36}$$

$$E_i = \frac{1}{M} \sum_{j=1}^{M} |c_{ij}|^2 \tag{6.37}$$

$$sd_i = \left(\sum_{j=1}^{M} \frac{(c_{ij} - vm_i)^2}{M-1} \right)^{1/2} \tag{6.38}$$

$$rms_i = \frac{1}{M} \sum_{j=1}^{M} |c_{ij} - vm_i| \tag{6.39}$$

Feature vectors are conformed by taking as vector components with the values of the features corresponding to the terminal frequency channels. A feature vector corresponding to each horizontal, vertical, and diagonal detail coefficients is generated.

This proposal is used in the latex drying experiment previously detailed, where the feature vector corresponding to each of the THSP images (texture patterns) were constructed. A set of 100 images samples was randomly extracted from each THSP image corresponding to each drying stage. Each image was decomposed (using the 2-D WT) up to a level $N = 2$. The texture features were calculated within each of the seven terminal-frequency channels for each image sample. Then, the mean value of these vectors from the image samples of the THSP image was obtained and consequently a vector of seven dimensions was assembled that characterizes each one.

The Mahalanobis distance between the sample feature vector ($u_f, f = 1 \dots 4$) (corresponding to the driest stage) and the feature vectors (u_{tf}, where $t = 1 \dots 14$) of the other texture patterns, with covariance matrix cov_{tf}, were computed by means of the equation:

$$D_{tf} = (u_f - u_{tf})^T \, cov_{tf}^{-1} (u_f - u_{tf}) \tag{6.40}$$

It was demonstrated that the features that showed better discrimination were the modulus and the mean residual. Using both features to characterize the time evolution of the drying process, the Mahalanobis distance was calculated between each texture pattern (to be $t = 1, 2, ... ,13$) and the corresponding to the THSP with lowest activity $t = 14$. The behavior of the distance over time (using both selected features: modulus and mean residual) showed an almost linear steep descent in the initial region, followed by a plateau-like or slow-variation region. The distance-adjusted curve evolves according to the drying process theories.

6.12   THSP CHARACTERIZATION BY FRACTAL ANALYSIS

In this section the THSP characterization by fractal analysis using the wavelet transform is proposed, using two different approaches: the Hurst exponent and the multifractal formalism.

As can be observed in Figure 6.1, the THSP power density spectrum estimation shows a 1/f-like scaling. It is remarkable that (a) the spectrum associated with THSP series can be distinguished from "white noise" and that (b) this frequency behavior shows the existence of certain regulation that withdraws it from a suspected stochastic process. This 1/f-like scaling pattern could provide a new means to characterize the apparently erratic and aperiodic intensity fluctuations within the dynamic speckle series, detecting different levels of activity.

The 1/f power spectrum characterizes a variety of time series corresponding to very different empirical events such as currents in semiconductors, geophysical records, economic data, traffic flow rates, image texture, and heart rate variability. The fractional Brownian motion (fBm) model is associated with this kind of dynamics.[28]

Estimating the generalized Hurst (H) exponent, (named in honor of Harold Hurst [1880–1978] and Ludwig Hölder [1859–1937] by Benoît Mandelbrot) for a data set provides a measure of whether the data is a pure random walk or has underlying trends. Another way to state this is that a random process with an underlying trend has some degree of autocorrelation. Under this paradigm, Passoni et al. modeled the THSP series to obtain a Hurst parameter-estimation-based descriptor.[29]

The Hurst parameter of a fractional Brownian motion (fBm) is seen as a parameter of the fractional Brownian motion. Mandelbrot and van Ness (1968) defined a family of processes they called fractional Brownian motions (fBm), characterized by stationary, Gaussian-distributed increments, whereas the time series is not stationary.[30] The normalized family of these Gaussian processes, B^H, is the one with $B^H(0) = 0$, almost surely zero mean, and its autocorrelation function is given by

$$R_{B_H}(t,s) = \mathrm{E}\{B_H(t)B_H(s)\} = \frac{\sigma^2}{2}[|t|^{2H} + |s|^{2H} - |t-s|^{2H}] \qquad (6.41)$$

for $s,t \in \Re$.

H represents the scaling exponent of the series and can be any real number in the range $0 < H < 1$. The points of fractal analysis are to test whether this scaling law holds for real-world time series, estimating the scaling exponent. Ordinary Brownian motion corresponds to the special case $H = 0.5$ and represents the boundary between antipersistent ($H < 0.5$) and persistent fBms ($H > 0.5$).

The fBm power spectral density analysis is widely used method for assessing the fractal properties of a time series. It is based on the periodogram obtained by the fast Fourier transform algorithm.

$$S_{B^H}(f) = \frac{\sigma^2}{|f|^{2H+1}}$$
(6.42)

where f is the frequency and $S_B(f)$ the corresponding squared amplitude. It is not a valid power spectrum in the theory of stationary processes because it is a noninte-grable function, but it could be considered as a generalized spectrum.[30]

Previous analysis have shown that, nevertheless, fBm is nonstationary, the wavelet coefficients outline a stationary process at each scale, and consequently wavelet analysis is well suited to fBm.[31–32] Also, it was stated that the self-similarity characteristic of fBm is replicated in its wavelet coefficients, whose variance changes as a function of scale, j, according to the power law

$$\log_2\left\{\frac{1}{N}\sum_j c_j^2\right\} = (2H+1)j + constant$$
(6.43)

An estimation of the parameter H can be obtained in two steps:

- Estimating the variance of the wavelet coefficients using

$$\log_2(var(c_j[n])) = \log_2\left(\frac{1}{n_j}\sum_{n=1}^{n_j} c_j^2[n]\right)$$
(6.44)

- Plotting $\log_2 (var[c_j])$ versus j and fitting a minimum square line. From the slope of the line, the estimation of H is obtained.

In particular, these properties are widely used for estimating H or the related spectral exponent to characterize dynamic speckle time series.[29] The performance of the descriptor characterized regions with different levels of activity, in experiments like the detection of painting drying time, the corn seed viability test, and the apple bruised area. Zunino et al. modeled the laser propagation through turbulent media considering a generalization of the former method where the parameter H is no longer constant, but it is a continuous function of time.[33]

Other approach based in a multifractal formalism was proposed by Federico and Kaufmann[34] to characterize the dynamic speckle phenomenon. Such analysis was carried out by calculating the Hausdorff dimension of the set of singularities of Hölder exponent. The singularities are determined by the wavelet transform modulus maxima of the intensity along THSP. It is stated that the properties of the dynamic speckle can be better characterized by using a set of scaling exponents describing the singularities of the intensity signal than by using the theory of fractional Brownian motion or frequency filter analysis. Hence, singularities and irregular structures of the scattered intensity signal carry the essential information of the underlined physical phenomenon, and multifractal analysis is well adapted to describe the statistical properties of dynamic speckle. Degaudenzi and Arizmendi[35] proposed a methodology for the multifractal analysis carried out by calculating the Hausdorff dimension of electrical owner demand patterns.

WT provides a useful tool for the detection of self-similarity or self-affinity in a temporal series. For a value t_0, the modulus of the transform is maximized when the frequency a is of the same order of the characteristic frequency of the signal $s(t)$ in the neighborhood of t_0. There it will have a local singularity exponent $h(t_0)$, which means that around t_0

$$|s(t) - P_n(t)| \approx |t - t_0|^{h(t_0)} \tag{6.45}$$

Where $P_n(t)$ is an n-order polynomial, and

$$W\psi(a, t_0) \approx a^{h(t_0)} \tag{6.46}$$

provided that the first $n + 1$ moments of ψ are zero.

Therefore, the WT can be used to analyze the local regularity of $s(t)$. The exponent $h(t_0)$ measures how irregular the signal is at the point t_0. The singular spectrum $D(h)$ of $s(t)$ is the function that associates to any exponent h the Hausdorff dimension of the set of all the points t_0 such that the Hölder exponent of s at t_0 is $h(t_0)$. To compute $D(h)$, this multifractal approach takes advantage of the time-scale partitioning given by the maxima representation to define a partition function $Z(a, q)$, which is expressed in the limit $a \rightarrow 0+$ as

$$Z(a, q) = \sum_{\{t_i(a)\}} |W_\psi s(a, t_i(a)|^q \approx a^{\tau(q)} \tag{6.47}$$

where $\{t_i(a)\}$ are the chains of the wavelet maxima across the scales, and $q \in \mathfrak{R}$. The most pronounced modulus maxima take place when very deep singularities are detected, whereas the others correspond to smoother singularities. Both functions $D(h)$ and $\tau(q)$ describe the same aspects of a multifractal, and they are related to each other. In fact, the relationships are

$$\tau(q) = D(h) - qh \tag{6.48}$$

Where h is given as a function of q by the solution of the equation

$$\frac{d}{dh}(qh - D(h)) = 0 \tag{6.49}$$

These two equations represent a Legendre transform from the variables q and τ to the variables h and D.

Analyzing two THSP series corresponding to the coin experiment detailed in Chapter 5, one belonging to different coating thickness by computing the scaling exponent $\tau(q)$ and the singularity spectrum $D(h)$, the associated spectrum $D(h)$ displayed the commonly observed bell shape (Figure 6.16 and Figure 6.17). The spectrum $D(h)$ is a convex function and its maximum is the fractal dimension of

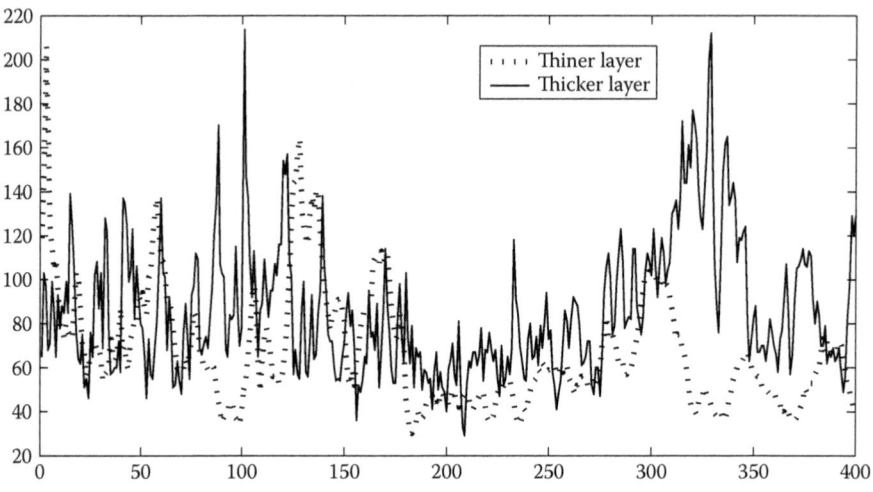

FIGURE 6.16 THSP time series corresponding to different coating thickness of wet paint.

the exponent $h(D_{max})$ most frequently encountered in the temporal sequences corresponding to the analyzed pixel. If the distribution is homogeneous, there will be an unique $h(D_{max}) = H$ (Hurst exponent); but if it is not, there will be several exponents h like in this situation. The most frequent h will characterize the series and will play as Hurst exponent, characterizing the activity of the scatterers, although other ones can be selected depending on the involved process.

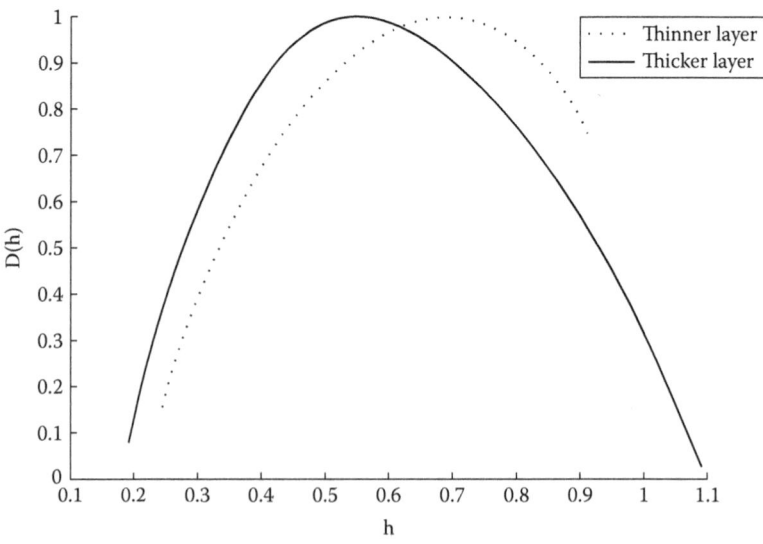

FIGURE 6.17 $D(h)$ spectrum of two different coating thickness of wet paint.

This latter approach, the multifractal one, has proved to be richer than the single H estimator, given that the $D(h)$ spectrum provides a set of parameters inherently useful to feature the time series like

$$\bar{h} = (h_{min} + h_{max})/2 \qquad\qquad (6.50)$$

$$h_{range} = (h_{max} - h_{min}) \qquad\qquad (6.51)$$

As can be evaluated, the bell-shaped spectrum functions present well-differentiated parameters useful to characterize the different topographic areas of the painted coin, enabling to ensemble a similar image to the one shown in Chapter 5.

Although the multifractal formalism has proven to be useful to characterize THSP patterns, it requires a certainly high computational cost compared to other processes such as those that deal with time values.

REFERENCES

1. Oppenheim, A. V., Schafer, R. W., and Buck, J., *Discrete-Time Signal Processing*, Second edition, Prentice Hall, Englewood Cliffs, 1999.
2. Manolakis, D. G., Ingle, V. K. and Kogon, S. M., *Statistical and Adaptive Signal Processing*, Artech House, Boston-London, 2005.
3. Dubois, E., The sampling and reconstruction of time-varying imagery with application in video systems, *IEEE Proc.*, 73: 502, 1985.
4. Aizu, Y. and Asakura, T., Bio-speckles, in *Trends in Optics*, Ed. A. Consortini, London, Academic Press, 1996, 27–49.
5. Ellis, M. G., *Electronic Filter Analysis and Synthesis*, Artech House, Boston-London, 1994.
6. Stoica, P. and Moses, R., *Introduction to Spectral Analysis*. Prentice Hall, Englewood Cliffs, 1997.
7. Gabor, D., Theory of communication, *J. Inst. Elect. Eng.*, 93(26), 429, 1946.
8. Morlet, J. et al., Wave propagation and sampling theory, *Geophys.*, 47, 203, 1982.
9. Daubechies, I., Where do wavelets come from? A personal point of view, *IEEE Proc.*, 84, 510, 1996.
10. Daubechies, I., *Ten Lectures on Wavelets*. Philadelphia: Society for Industrial Mathematics (SIAM), 1992.
11. Phillips, W. J., Time-Scale Analysis. Wavelets and Filter Banks Course Notes, http://www.engmath.dal.ca/courses/engm6610/notes/node4.html (accessed October 20, 2007), 2003.
12. Sendra, G. H. et al., Decomposition of biospeckle images in temporary spectral bands, *Opt. Letters*, 30(13),1641, 2005.
13. Sendra, G. H. et al., Biological specimens analysis using dynamic speckle spectral bands, *Proc. SPIE*, 5858, 287, 2005.
14. Murialdo, S. et al., Application of a laser speckle method for determining chemotactic responses of Pseudomonas aeruginosa towards attractants. In *Speckle '06 From Grains to Flowers,* Slangen, P. and Cerruti, C., Eds., *Proc. SPIE*, 2006, 6341 2D.
15. Fujii, H. and Asakura, T., Blood flow observed by time-varying laser speckle, *Opt. Lett.* 10(3), 104, 1985.
16. Aizu, Y. and Asakura, T., Bio-speckle phenomena and their application to the evaluation of blood flow, *Opt. Laser Tech.*, 23 (4), 205, 1991.

17. Ruth, R., Haina, D., and Waidelich, W., The determination of a blood flow parameter by the Doppler method, *Opt. Acta*, 31, 759, 1984.

18. Ruth, B., Superposition of two dynamic speckle patterns: An application to noncontact blood flow measurements, *J. Mod. Opt*, 34, 257, 1987.

19. Shannon, C., A mathematical theory of communication, *Bell System Tech. J.*, 27, 379, 1948.

20. Powell, C. E. and Percival, I. C., A spectral entropy method for distinguishing regular and irregular motion of Hamiltonian system, *J. Phys. A.*, 12, 2053, 1979.

21. Blanco, S. et al., Time-frequency analysis of electroencephalogram series (III): Wavelet packets and information cost function, *Phys Rev. E.*, 57: 932, 1998.

22. Rosso, O. et al., Wavelet entropy: A new tool for analysis of short duration brain electrical signals, *J. Neurosc. Meth.*, 105, 65, 2001.

23. Passoni, L. I. et al., Dynamic speckle processing using wavelets base entropy, *Opt. Comm.*, 246, 219, 2005.

24. Oulamara, A., Tribillon, G., and Duvernoy, J., Biological activity measurements on botanical specimen surfaces using a temporal decorrelation effect of laser speckle, *J. Mod. Opt.*, 36, 165, 1989.

25. Amalvy, J. I. et al., Application of dynamic speckle interferometry to the drying of coatings, *Prog. in Org. Coat.*, 42, 89, 2001.

26. Braga, R. A. et al., Biological feature isolation by wavelets in biospeckle laser images. *Comp. Electr. Agric.*, 58, 123, 2007.

27. Limia, M. F. et al., Wavelets transform analysis of dynamic speckle patterns texture, *Appl. Opt.*, 41, 6745, 2002.

28. Mandelbrot, B. B. and Ness, J. W. V., Fractional Brownian motions, fractional noises and applications, *SIAM Rev.*, 4, 422, 1968.

29. Passoni, L. I., Rabal, H., and Arizmendi, C. M., Characterizing dynamic speckle time series with the Hurst coefficient concept, *Fractals*, 12, 319, 2004.

30. Peréz, A. et al., Analyzing blood cell concentration as a stochastic process, *IEEE Eng. Med. Biol. Mag.*, 20(6), 170, 2001.

31. Flandrin, P., On the spectrum of fractional Brownian motions, *IEEE Trans. Inf. Theory*, 35, 197, 1989.

32. Flandrin, P., Wavelet analysis and synthesis of fractional Brownian motion, *IEEE Trans. Inf. Theory*, 38, 910, 1992.

33. Zunino, L. et al., Characterization of laser propagation through turbulent media by quantifiers based on the wavelet transform: Dynamic Study, *Phys. A*, 364, 79, 2006.

34. Federico, A. and Kaufmann, G. H., Multiscale analysis of the intensity fluctuation in a time series of dynamic speckle patterns, *Appl. Opt.*, 46, 1979, 2007.

35. Degaudenzi, M. E. and Arizmendi, C. M., Wavelet-based fractal analysis of electrical owner demand, *Fractals*, 8, 239, 2000.

7 Granular Computing in THSP Fuzzy Granular Analysis

Ana Lucía Dai Pra and Lucía Isabel Passoni

CONTENTS

7.1 INTRODUCTION

Granular computing is proposed as a new approach for activity detection in dynamic speckle phenomenon. Once laser speckle appearance is intrinsically granular, it is possible to suggest a similarity with the basis of the new granular computing (GrC) paradigm. The granulation is a human view, whereas the grouping of some entities (colors, numbers, pixels of images, shapes, etc.) facilitates the perception of the surrounding physical world. Likewise, the information about granularity is a way of abstraction to simplify the information complexity, grouping the information in blocks that can be used for classification, modeling, prediction, and interpretation. In this way, subsets of big and complex data sets can be handled maintaining a whole integrity.

Granular computing term dates from 1970s, when it was proposed by Zadeh in his early works on fuzzy sets.[1] Then, the term has been extended to other fields and fast growing interest in GrC has arisen. It plays important roles in bioinformatics, e-business, security, machine learning, data mining, and high-performance computing. The GrC is originated from different information processing fields, such as artificial intelligence, theoretical computer science, interval computing, cluster analysis,

169

fuzzy and rough set theories, belief functions, machine learning, and databases. In the last years, GrC has begun to play an important role as a general computation theory for effectively using granules such as classes, clusters, subsets, groups, and intervals to build an efficient computational model for complex applications with a great quantity of data, information, and knowledge.[2-3]

7.2 INFORMATION GRANULES

Zadeh, in his earlier GrC definitions, proposed the information representation as a number of entities or information granules, viewed as linked collections of objects (data points, in particular) drawn together by the criteria of indistinguishability, similarity, or functionality. A representation of granules must capture their essential features and make explicit a particular aspect of their physical meaning.[4] Information granules can be related to any information type within any context.

7.2.1 TEMPORAL GRANULARITY

The aim of temporal granularity is the transformation and reduction of signals to sets of information granules to facilitate their analysis, interpretation, or classification. This transformation converts a time series into a set of temporal granules generated on time intervals, changing the notion of the traditional signal processing.[5]

Given the intrinsic dynamic characteristic of the laser speckle patterns, the processing of THSP signals with temporal GrC is proposed in this chapter. The granular transformation of THSP is considered a convenient approach to deal with these complex signals.

7.3 FRAMEWORK

Considering the numeric nature of signals or time series, there are a number of formal frameworks in which information granules can be created. Basically, we will be considering

- Set theory and interval analysis
- Fuzzy sets
- Rough sets and shadowed sets

These are well-known, thoroughly investigated, and come with a vast array of applications. The fuzzy, rough, and shadowed sets are extensions of the classical set theory that permits consideration of the uncertainty information.

7.3.1 SET THEORY AND INTERVAL ANALYSIS

Considering the classical set theory, given a set of objects X whose elements are denoted x_i and a classical subset A of X, the membership of each x_i to A is viewed as a characteristic function μ_A from X to $\{0,1\}$ such that:

$$\mu(x_i) = \begin{cases} 1 \textit{ iff } x_i \in A \\ 0 \textit{ iff } x_i \notin A \end{cases} \qquad (7.1)$$

In general, numerical data sets are related to numerical subintervals. For example, in the pixel intensity values set, subsets could be characterized by arbitrary terms such as dark, medium, and light, each term corresponding to numeric subintervals in the integer number interval [0,255].

In many cases, the subintervals definition for some terms or subsets are not mathematically defined or clearly bounded; they are subjective terms and can be partially overlapped. This problem has led to the study of other types of sets that permit classifying objects encountered in the real physical world. The fuzzy, rough, and shadowed sets are examples.

7.3.2 FUZZY SETS

Fuzzy sets facilitate the interpretation of numerical interval with undefined bounds. These sets are characterized by a membership function that assigns to each set element a grade of membership ranging between zero and one. Given a set of objects X whose elements are denoted x_i and a subset A of X, the membership of each x_i to A is viewed as a characteristic function μ_A from X to real interval [0,1]. A is named a fuzzy set and $\mu_A(x_i)$ is the grade of membership of x_i in A. A is a subset of X that has no sharp boundary and is characterized by the set of pairs:[6]

$$A = \{(x_i, \mu_A(x_i)), x_i \in X\} \qquad (7.2)$$

The notion of fuzzy sets provides a convenient point of departure for the construction of a conceptual framework which parallels in many aspects the framework used in the case of ordinary sets, but with a much wider scope of applicability, particularly in the fields of pattern classification and information processing. Essentially, such framework provides a natural way of dealing with problems in which there are no defined criteria of classification or membership.[7]

Several types of different functions can be used to define fuzzy sets membership; the characteristic of all of them is that they take gradual values in the interval boundaries. The membership values indicate a certainty value about a term or concept that can be overlapped to other ones. An example is shown in Figure 7.1.

Fuzzy sets exhibit a well-defined semantics and emerge as fully meaningful and conceptual entity building modules identified in problem solving.[8] Their application in granular computation helps interpret the granule abstract concept.

7.3.3 ROUGH SETS AND SHADOWED SETS

Rough and shadowed sets are based on fuzzy sets. The shadowed sets are induced from fuzzy sets by accepting a specific threshold level for the membership functions.[9]

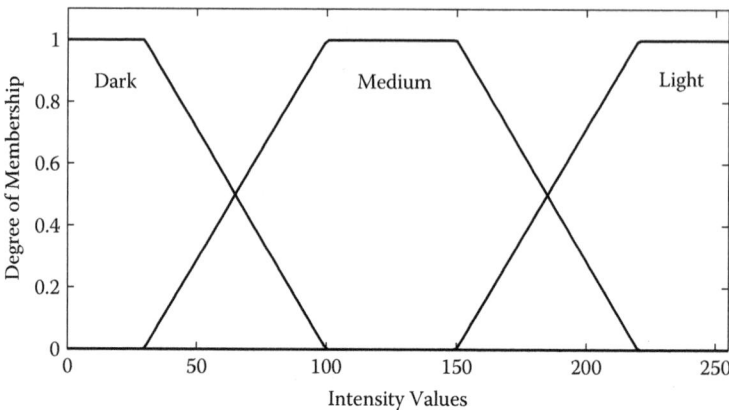

FIGURE 7.1 Membership functions.

The rough sets are induced from fuzzy sets by accepting a specific similarity level for the membership functions, defining lower and upper approximations of a fuzzy set.[10]

Both classes of sets (level fuzzy sets) could be used in the present application to simplify the computation in specific speckle activity classification.

7.4 THSP FUZZY TEMPORAL GRANULARITY

The temporal granularity in THSP is analyzed with the fuzzy sets concept. Distinct examples of fuzzy temporal granulation can be found in the literature, applied to find similarity between time series;[11] describe time series, such as ECG signals,[5,12] and time series analysis.[13]

The granular computing is presented as a novel methodology, which converts the THSP signal in a set of information granules. The selected criterion for the discernment of the temporal granules is to classify the intensity value in three fuzzy sets: dark, medium, and light. These fuzzy sets are overlapped, which means that some intensity values are light in some grade and simultaneously medium in another grade. In the same way, some medium intensity values are simultaneously dark in some grade.

The fuzzy membership functions with respect to the mentioned overlapping linguistic concepts (dark, medium, and light) may be defined for each fuzzy set. Three fuzzy sets with trapezoidal functions membership are selected, and the function parameters are established according to an equalization criterion.

The fuzzy membership function parameters are defined through observation of the intensity histogram of a speckle pattern image, given that the intensity distribution does not show any noticeable variations along the experiment.

A histogram with 256 bins, corresponding to each i intensity values in [0,255], shows the distribution of pixels intensity values of the speckle pattern. The interval [0,255] can be partitioned in five sectors, three corresponding to fuzzy membership function equal to 1, and two intermediate sectors with increased or decreased values in the real interval [0,1], defining the three trapezoidal functions. (See Figure 7.2.)

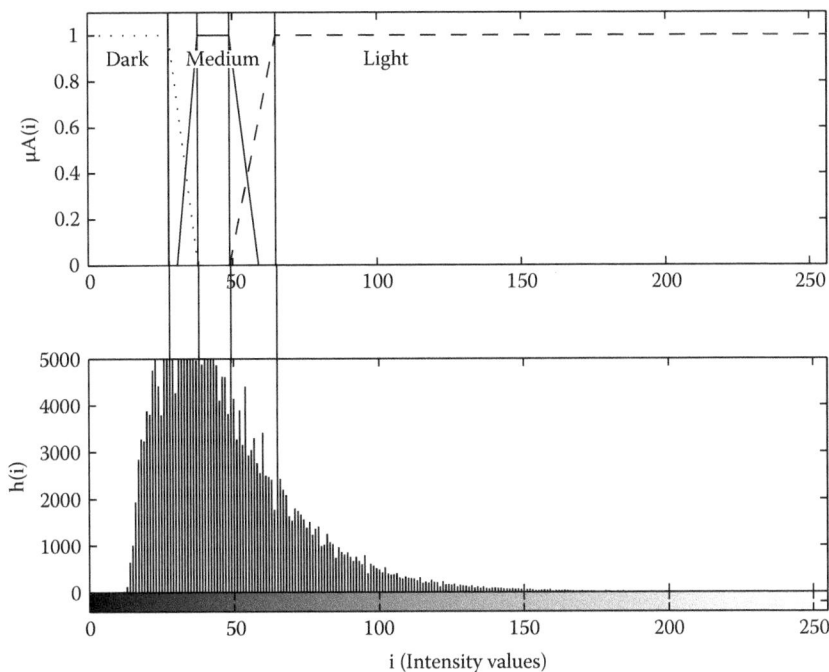

FIGURE 7.2 Image histogram and corresponding fuzzy membership functions.

Given the $h(i)$ values, the quantity in each bin, and $\mu_k(i)$ the trapezoidal function applied to each i value, it is expected that the $\mu_k(i)$ fulfills the following proposed condition:

$$\sum_i \mu_{\text{dark}}(i) \cdot h(i) \cong \sum_i \mu_{\text{medium}}(i) \cdot h(i) \cong \sum_i \mu_{\text{light}}(i) \cdot h(i) \qquad (7.3)$$

Another methods based on fuzzy entropy function[14] can be used to find the adequate parameters of the membership functions, or to justify if the information contained in each fuzzy data set is similar. Different techniques for intensity values fuzzy partition based on fuzzy entropies are found in the literature.[15]

This methodology enables us to find the intervals optimized according to the experiment. Consequently, these adjustments will eliminate undesirable effects caused by the illumination and contrast variations among different essays.

Once the fuzzy membership functions are defined, they must be applied to each pixel in the speckle pattern sequence. Given a sequence of T intensity values; and k fuzzy sets defined in the universe of grey level values; the quantity of granules Q_T in T samples is computed as:

$$Q_T = \frac{\left(\sum_{'k} \sum_{i=2}^{T} seq_{i,k}(\mu_k(x_i)) \right)}{T} \qquad (7.4)$$

With

$$seq_{i,k}(\mu_k(x_i)) = \begin{cases} 1 & \text{if } \mu_k(x_{i-1}) \neq 0 \quad \text{and} \quad \mu_k(x_i) = 0 \\ 0 & \text{in other case} \end{cases} \tag{7.5}$$

seq i,k(.) is 1 when a sequence of values greater than zero corresponding to the same fuzzy set finishes, determining one information granule.

In order to determine an index activity, only the relative number of eventually overlapped granules Q_T is considered. Q_T is the activity index for a determined THSP pixel.

Figure 7.3 shows a visualization of the methodology applied to an arbitrary signal. On the left of the grey intensities axis, the fuzzy membership functions are presented. Under the signal, the corresponding dark, medium, and light granules are shown in their corresponding intensity scales. Finally, the successive resultant Q_T values are represented, each one describing the former activity as a sequential process.

For each pixel, a relative temporal quantization of overlapped granules is computed, differing from some clipping methods that require crossing statistics at a single level.[17] The activity can be obtained in each instance, being almost a real-time process.

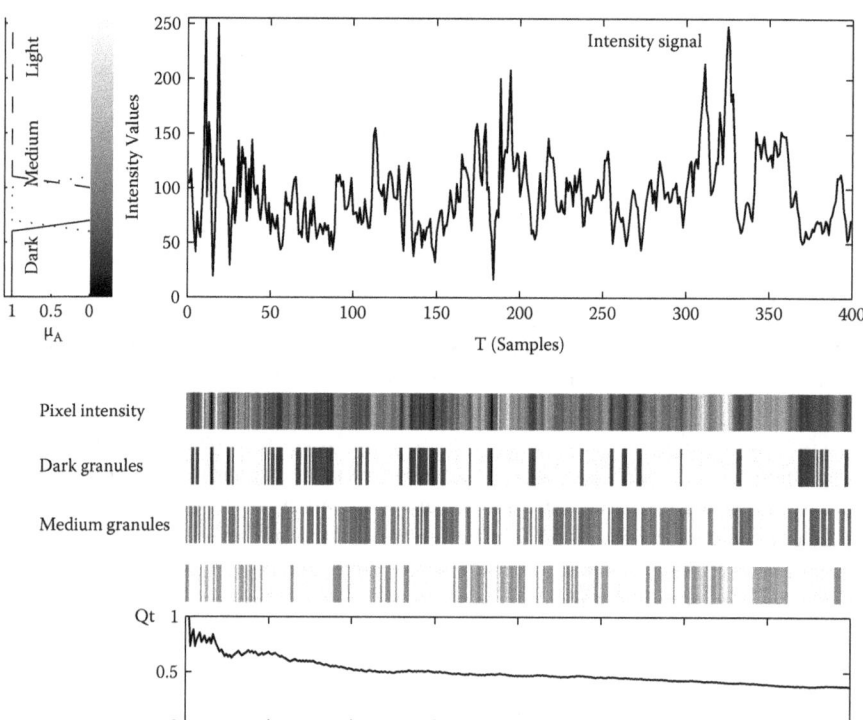

FIGURE 7.3 Signal granulation process.

The quantity of fuzzy sets can be modified; in general, their increase produces no improvement, and the activity images tend to be noisy. A α-cut can be applied to the fuzzy sets to decrease the overlapping; also, it generally produces noisy activity images.[6]

Consequently the process of quantification of the fuzzy granules becomes in an indicator of the activity of the THSP. Other ways to characterize the THSP could be based on the distrubution and features of the different granules that compose it (dark, medium and light). By means of this abstraction it is possible to consider new processes of analysis of complex signals.

7.5 SOME RESULTS

In others chapters, the successful evaluation of coatings dried using dynamic speckle was presented. Hence, it is proposed to discover topographical information under a fresh paint layer. A deposition of a thin layer of fresh paint was performed on an object with known surface topography (a coin). Places where the paint layer was thinner were expected to be drier than those where the layer was thicker and consequently less active.

Figure 7.4 shows a speckle pattern image where the topography is not identifiable. The histogram of intensity values distribution of the speckle pattern image and the fuzzy membership functions obtained starting from the intensity values distribution are shown in Figure 7.2. Figure 7.5 shows a progressive sequence of activity images started from the first speckle pattern. The number under each subimage indicates the quantity of former speckle patterns that was evaluated. Starting from the 15th pattern, only some results are shown to avoid repetitive images. It can be observed that the activity image is discovered in a few samples. Finally, Figure 7.6 shows the activity image obtained with 200 samples.

In sequence, the results of some experiments already presented in other chapters are shown here. Figure 7.7 corresponds to the activity image for a corn seed viability

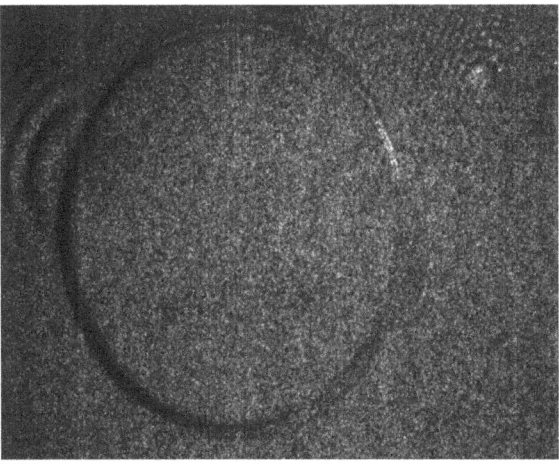

FIGURE 7.4 Painted coin speckle pattern image.

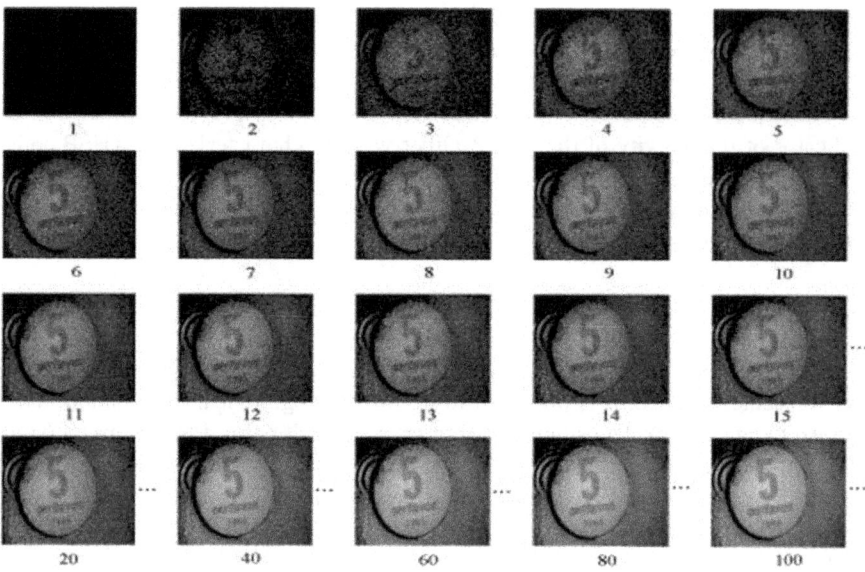

FIGURE 7.5 Sequence of activity images.

test. Figure 7.8 corresponds to the activity image for a bruised apple. Figure 7.9 shows the activity graph for a nail enamel drying process; in this case the graph represents the mean activity in each THSP pattern column. Finally, Figure 7.10 shows the activity graph for the drying process of latex paint. These represent the mean activity in each THSP pattern.

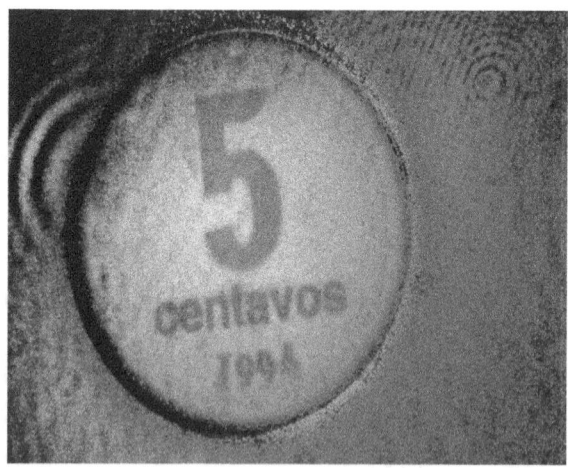

FIGURE 7.6 Final painted coin activity image.

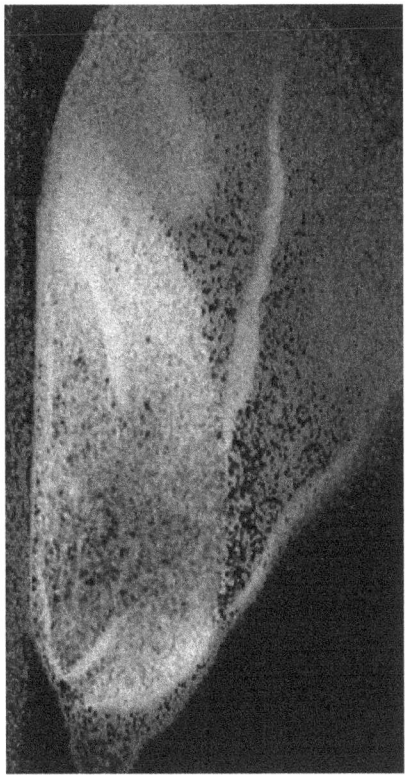

FIGURE 7.7 Activity image for a corn seed viability test.

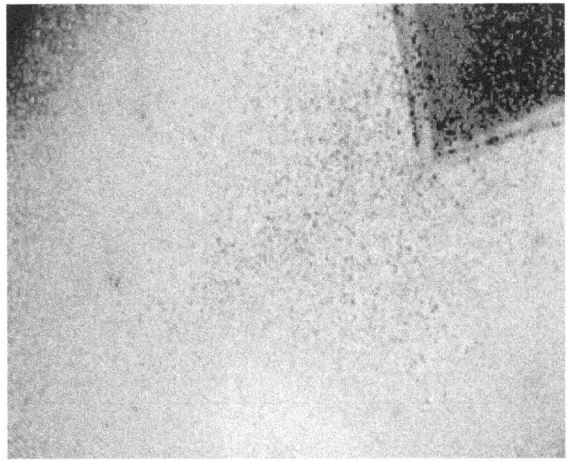

FIGURE 7.8 Activity image of apple bruising.

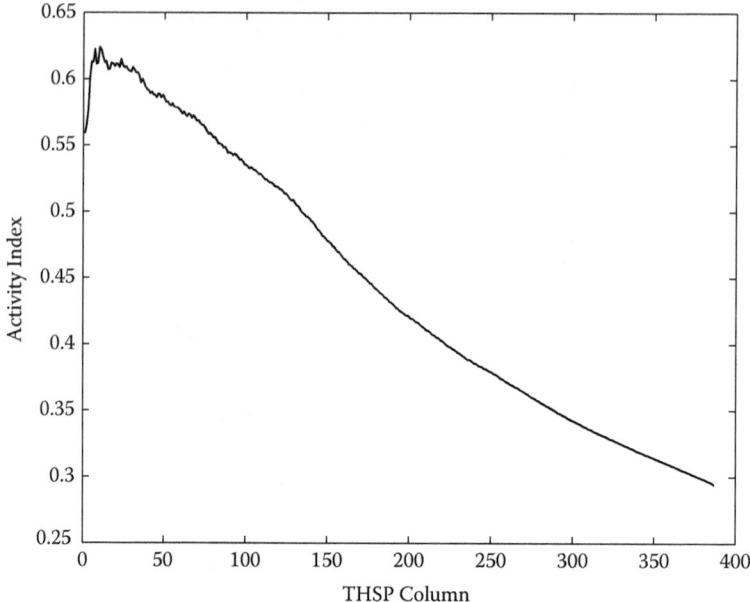

FIGURE 7.9 Activity index for the nail enamel drying process.

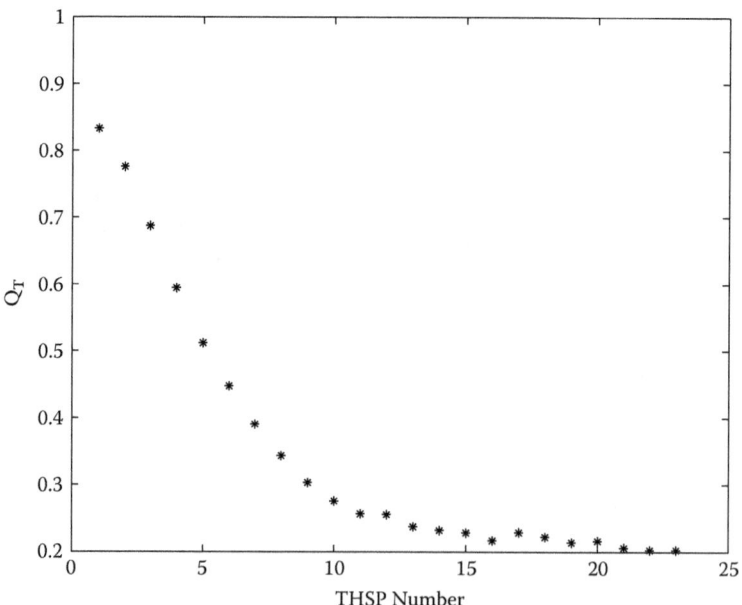

FIGURE 7.10 Activity index for the latex paint drying process.

7. 6 CONCLUSIONS

The temporal granulation process is applicable to the determination of activity identifiers in THSP patterns. This methodology gives us the opportunity to characterize or compare activities through the analysis of THSP granulated signals, evaluating the granule characteristics such as size, distribution, and fuzzy weight. The processing involves low computational cost and can be used with smaller amount of data than other methods. Furthermore, it is less sensitive to noise, does not necessarily require the setting of definite time intervals, and can be implemented in almost real time.

REFERENCES

1. Zadeh, L. A., Toward a theory of fuzzy information granulation and its centrality in human reasoning and fuzzy logic, *Fuzzy Set. Syst.*, 90, 111, 1970.
2. Bargiela, A. and Pedrycz, W., *Granular Computing: An introduction*, Kluwer Academic Publishers, Boston, 2003.
3. Bargiela, A. and Pedrycz, W., The roots of granular computing, in *Granular Computing, IEEE International Conference Proceeding*, IEEE Press, New York, 2006, 806.
4. Yao, Y., The art of granular computing, in *Rough Sets and Emerging Intelligent Systems Paradigms*, Kryszkiewicz, M., et al., Eds., Springer, Berlin, 2007, 101.
5. Pedrycz, W., Temporal granulation and its application to signal analysis, *Inform. Sci.*, 143, 47, 2002.
6. Dubois, D. and Prade, H., *Fuzzy Sets and System: Theory and Applications,* Academic Press, New York,1980.
7. Zadeh, L. A., Fuzzy sets, *Inform. Control*, 8, 338, 1965.
8. Pedrycz, W., *Granular Computing: An Emerging Paradigm*, Phisica-Verlag, New York, 2001.
9. Pedrycz, W., Shadowed sets: Representing and processing fuzzy sets*, IEEE Trans. Syst., Man Cybern.*, 28, 103, 1998.
10. Pawlak, Z., Rough sets: A new approach to vagueness, in *Fuzzy Logic for the Management of Uncertainty*, Zadeh, L. and Kacprzyk, J., Eds.,105, John Wiley, New York, 1992, 105.
11. Yu, F., Chen, F., and Dong, K., A granulation-based method for finding similarity between time series, in *Granular Computing, IEEE International Conference Proceeding*, IEEE Press, New York, 2005, 700.
12. Gacek, A. and Pedrycz, W., A Granular description of ECG signals, *IEEE Trans. Biomed. Eng.*, 53, 1972, 2006.
13. Bargiela, A. and Pedrycz, W., Granulation of temporal data: A global view on time series, in *Fuzzy Information Processing Society, NAFIPS Proceeding*, IEEE Press, New York, 2003, 191.
14. De Luca, A. and Termini, S., A definition of nonprobabilistic entropy in the setting of fuzzy sets theory, *Inform. Control.*, 20, 301, 1972.
15. Zhao, M., et al., A technique of three-level thresholding based on probability partition and fuzzy 3-partition, *IEEE Trans. Fuzzy Syst.*, 9, 469, 2001.
16. Pal, S. K. and Rosenfeld, A., Image enhancement and thresholding by optimization of fuzzy compactness, in *Fuzzy models for pattern recognition*, Bezdek, J. C. and Pal,. S. K., Eds., 369 IEEE Press, New York, 1992, 369.
17. Brown, J., Jakeman, E., and McWhirter, J. G., Clip-level crossing rates in photon correlation and analogue signal processing, *J. Phys. A: Math. Gen.*, 10, 1791, 1977.

8 Applications in Biological Samples

Roberto A. Braga Jr., Giovanni F. Rabelo,
Joao Bosco Barreto Filho, Flávio M. Borém,
Joelma Pereira, Mikiya Muramatsu,
and Inácio Maria Dal Fabbro

CONTENTS

8.1 INTRODUCTION

Applied science has been a branch of research attracting specialists searching for reliable solutions to problems in many areas. The challenging developments in laser technology have also attracted researchers, particularly physicists, engineers and biologists from many disciplines, to the specialized area of the dynamic speckle technique as a tool to monitor biological attributes. The key word featured in all applications of the dynamic speckle research line in biological tissues is *activity*. Dynamic speckle, in all of its applications, involves the need for measurements with distinct approaches, as presented in the preceding chapters, since the changes in the speckle pattern are produced by a large range of interactions between coherent light and complex biological material. Work with biological samples has brought new features under analysis, the most prominent involving variability. This means that it is impossible to get the same answer with precision—in other words, with repeatability—even when illuminating specimens with the same expected characteristics. The variability of a biological sample can be managed with the use of replications and statistical analysis to evaluate the variance level of the results. Therefore, reports on this kind of application have mostly been based on a strict evaluation of mean values and other statistical measures.

Besides variability, biological samples are very complex having a net of interactions associated with a large number of elements. Common sense allows the conclusion that the results presented by dynamic speckle are related to many factors in a collective phenomenon by means of a global measure, which offers a challenge to researchers to circumvent many obstacles between the observed measure and the desired variable. Nevertheless, the global measure could be a better way to get a robust result, as will be discussed in some applications. The reader should realize that each case may produce different answers under different dynamic speckle analysis methods, therefore an unsuccessful result should be faced as a challenge, and not as an end point of investigations. Consequently, a good result cannot be considered as the only explanation for the whole process. Finally, the reader should also realize that we are observing the sample from one point of view, in a new dimension in which the results are shown as indirect information of the whole activity of the material and its complex interaction with the laser beam.

This chapter covers dynamic speckle applications in biological areas, presenting them as found in agriculture, medicine, food, and elsewhere. The reader will be presented with a number of terms employed in this technique, such as laser speckle imaging (LSI), biospeckle laser (BSL), endoscopic laser speckle imaging (eLSPI), and laser speckle contrast (LASCA). It is, of course, not possible to cover all the dynamic speckle applications for there a thousands of claimed processes. A brief review of the interaction of light with matter will be given, followed by typical approaches adopted to obtain information from illuminated objects.

8.2 DATA ACQUISITION SYSTEM

The experimental design of dynamic laser speckle in biological systems are based on both analog and digital devices. Analog observations started in the 1970s with the use of photo-detectors, which was the most developed technology of those years.[1] Subsequent improvements in electronics and digital devices have given rise to a new generation of detectors using charge-coupled devices (CCDs)—though photodiodes provide, in most cases, a faster and more sensitive answer than the CCDs. The observer, represented by the light detector, plays a special role in speckle monitoring.

Experimental design also depends on the application and on which features need to be measured. Two approaches appear in most applications, and they are related to how light collects information from tissues. They are

- *Back-scattering speckle*
- *Forward-scattering speckle*

The best way to introduce the interaction of light and the detector is to review some light and optical concepts, and their interaction with biological matter.

Electric field in every point of a monochromatic light beam with amplitude (*a*), wave length (λ), frequency (ω), and propagating in a (*z*) direction can be expressed by:

$$E\,(z,t) = \text{a.cos}(\omega t - k.z) \tag{8.1}$$

where,

$E(z,t)$—electric field
$\omega = 2.\pi.v$—angular frequency
$K = 2.\ \pi/\lambda$—wave number
λ—wavelength
t—time
z—space

In its complex representation with exponential function, it can be expressed as:

$$E(z,t) = \ \text{Re}\{\text{a.exp}[(\,j.\omega.t - k.z)]\} \tag{8.2}$$

where,

Re{}—real part of the expression

The interaction of an electric field with matter is stronger than that of the magnetic field, so it is usual to consider only the electric field in the analysis. When we know the electric field, it is possible to obtain the magnetic one, therefore, is not necessary to treat both fields.

Even if the CCD camera is only sensitive to energy, statistics obtained from the speckle pattern depend on whether the light is polarized or nonpolarized. If there is no polarizer between the sample and the camera, each perpendicular state of polarization behaves as an independent speckle pattern and both are added on an intensity basis. The contrast is lower in this case than if there is a polarizer that permits only one state to be recorded.[2]

Detectors are sensitive to the light power when we are working with visible light. In other words, detectors do not respond to changes in the direction of the electric vector. Therefore, the measure is produced by the square of the amplitude of the field, which is the intensity or irradiance of the light. The energy, which passes through a unitary area in the orthogonal direction of propagation, is proportional to the temporal mean squared value of the field

$$I = (E^2)\lim_{T \to \infty} \int_{-T}^{T} E^2 dt = a^2/2 \tag{8.3}$$

where,

a—amplitude of the electric field
E—electric field
I—intensity of the electric field
t—time

In accordance with the Lambert, Bourguer, and Beer laws, there is a relationship between the quantity of incident radiation and the quantity absorbed in the light path. The radiation suffers a variation dI_λ, and can be expressed by

$$dI_\lambda = -\mu_\lambda \cdot I_\lambda \cdot dz \tag{8.4}$$

where,

z—thickness of the material
$\mu_\lambda \cdot I_\lambda$ —Lambert coefficient of absorption
λ—wavelength

When the speckle is formed by a portion of light that was scattered by moving elements, it is modulated by such movement. It is difficult to identify precisely which movement is the most relevant for the scattering, because laser penetrates into the biological material and experiences multiples deviations in its path, as well as losing polarization before returning to the surface and reaching the light detector. Such an effect may produce different sizes of speckle grain, depending on the path and the setup of the acquirement system. The grain size of a stationary speckle corresponds to the diameter of the diameter of the illuminated area,[3] expressed as:

$$d = 2.44 \left(\frac{\lambda \cdot z}{D} \right) \tag{8.5}$$

where,

d—speckle diameter
z—distance to the observation screen
D—diameter of the illuminated area

Therefore, the size of the speckle grains may be larger or smaller than the resolution of the detector, depending on the optics aperture. The speckle pattern is formed by two components: A set of big speckles originated in light scattered on the surface, which is dependent on the angle of the light, is modulated by small speckles produced by the light that emerges from inner portions of the material and is not dependent on the incidence angle.[4]

The experimental design involves several variables and, thus, when developing an experiment with similar samples, a protocol should be considered with respect to the distance and angle between samples and the detector, polarization and light intensity, and the size of the speckle laser pattern.

The intensity of the laser over the whole illuminated area is difficult to control due to its changing profile and the distribution of light over the surface. In most cases, laser light passes through the superficial layers and enters the sample. It is important to keep the depth reached by the illumination for all samples constant, so neutral filters can be used to control the intensity of the illumination. In addition, during experiments using biospeckle, some noise cannot be avoided. There are multiple sources of noise—for example, from the detector, environment, temperature, and humidity variations, among others. Noise is difficult to control, but in most cases it can be partially filtered during the image processing.

In order to control the movement of air and humidity, the experiments may be conducted in a closed environment, such as a glass box. Unfortunately, the noise coming from the detector is difficult to eliminate, when the CCD sensors show, in the absence of light, an image with some bright points—that is, a noise produced by thermal interference, because most detectors have no refrigeration system. Filtering this kind of noise can only be done in the image processing stage. However, the efficiency of filtering depends on the type of image processor, which must have a high quantization of 16 bits or more.

8.2.1 BACK-SCATTERING SPECKLE

Biospeckle images can be obtained from back-scattered light, when the laser beam reaches the surface of the specimen and is reflected, captured by a CCD camera, and transferred to the processing system (Figure 8.1). Incident light can also provide information about the internal structure of the material, giving to the scattered beam a composition of superficial and internal data. This is an important feature of back-scattered configuration, as the portion of light that is not absorbed by the inner structure brings useful information from the portion near to the surface.

It can be expected that if the power of the laser beam is able to reach deeply into the material and return, analysis of the dynamical behavior of inner particles is possible.

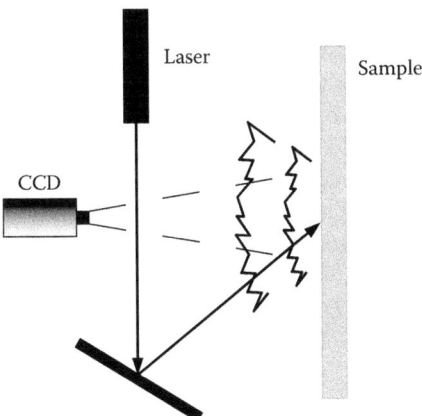

FIGURE 8.1 Arrangement setup with back and forward scattering.

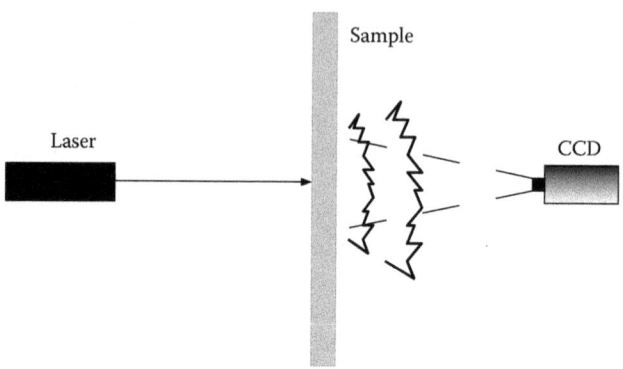

FIGURE 8.2 Arrangement setup with forward scattering.

Is this the only kind of interaction? The answer is no, because incident light interferes with both reflected and scattered light too. If the angle of incidence of the light is appropriately handled (by using a mirror, for example) the specular reflection can be avoided, whereas the diffuse reflection will interfere with the rays that come from the interior of the material to form the biospeckle.

8.2.2 FORWARD SPECKLE

Speckle can also be formed by the laser light that passes through a material, and in this case, the incident light is modulated by the scattered beam (Figure 8.2). The material therefore behaves as a very complex diffraction network with relative movements among the particles that direct the rays to form a moving grainy image. The movement can be presented in different levels of random composition, with the occurrence of absorption of light inside the material, so the level of that absorption depends on the sample's thickness.

We have seen that absorption of energy varies exponentially with the depth of penetration.[3] So, do we need more light to pass through the material compared to the back-scattering approach? If the material under study is very absorbent, the answer is yes, but if the material is transparent, we may expect that most of the light is not scattered and directly reaches the sensor, causing detector saturation (such as when we look directly to an object against the sun). This problem can be solved if we use a static diffuser before the light reaches the specimen or a diffuser between the CCD and the sample.

Some biological material like seeds, fruits, and foods are opaque, so the backscattering setup is more appropriate for speckle measuring. Some other biological specimens are diluted in fluids, a fact that can make the use of forward scattering easier. The diameter of the illuminated area can be expressed by Equation 8.6.

$$D = D_0 \left[1 - \exp\left(\frac{T}{T_0}\right) \right] \tag{8.6}$$

where,

D—diameter of the illuminated area

T—depth

T_0—1.6 mm

As mentioned in Chapter 2, $D_0 = 20$ mm and $T_0 = 1.6$ mm were empirically found in apples.

8.3 AGRICULTURAL APPLICATIONS

Due to the socioeconomic importance of guaranteeing quality food, there is a persistent search for the improvement of food products in terms of safety, health, appearance, and many other market attributes. Having the latest information about improved mechanical properties and assessing the stage of maturation of fruits and vegetables, for example, is important to the harvesting and postharvesting processes, as well as finding new methods of shelf life management of the products. At the present time, this information depends on both destructive and nondestructive tests which are hard to employ in an automatic production line, and in some cases is limited by tests with subjective human interpretations. The growing interest in machine based tests is focused on areas such as laser instrumentation with its ability to implement nondestructive and automated evaluations.

Besides food analyses such as those performed on fruits, seeds, biscuits, ice cream, and biological films, it is important to note the applications of dynamic laser speckle in animal reproduction in agricultural areas, which are thoroughly discussed in the following text.

8.3.1 FRUITS

Every biological material presents a structure composed by cells that are arranged to form a net of complex tissues. Vegetable cells, in particular, may vary in size, organization, and function before and after the harvest. The difference between vegetable and animal cells is the presence of cellular walls in the former. Such a wall is rigid, can reach 30nm in thickness, and may be divided into primary and secondary layers. The primary layer is formed by cellulose fibers, components that predominate inside the cell, wrapped in polysaccharide cement and proteins, which becomes rigid as the cells grow and are destroyed as the cells age. The mechanical support of the vegetable material is maintained by the cellular wall, and its basic function is to protect the cell's membrane against mechanical and osmotic ruptures. There are other elements that make up the architecture of the cellular wall—for example, hemicelluloses, pectin, and structural proteins.

After the harvest and during the maturity process, most of the pectin is dissolved, resulting in reduction of the attraction force between the cells. There are other parts of the cells that develop specific activities, but when we are interested in light propagation, the focus of the study should be the regulators of light penetration in the cells with the plastids called chloroplasts, chromoplasts, and leucoplasts assuming a special role. The distribution of chloroplasts in a tissue is not homogeneous, because they move to regulate the entrance of light, adjusting its absorption. This phenomenon, called cyclose, is important in the study of leaves. When a light beam reaches a vegetal tissue, it is partially reflected because the refractive index is larger than the air index. But due to the structural complexity of the vegetal tissue, the refractive index is different in every direction, so the light may be widely scattered and can suffer multiple specular and diffuse reflections.

When the speckle is formed by the portion of light scattered by movable elements, it is modulated by their movement. Then, it is difficult to identify precisely which element is responsible for the scattering, because a laser photo can penetrate the vegetal material and suffer multiple deviations in its path before it eventually returns to the surface and reaches the light detector.

Cellular structure may vary from one specimen to another, and the movement of cell components can also vary. This movement is also modified by the age of the cell. Therefore, it can be expected that the speckle formed by different cells is different; in addition, the speckle will change as the cell ages. In other words, a speckle can be used to distinguish specimens and also the age of biological material. We consider the age of a fruit as the number of days passed since the sample was brought from the mother plant. Temporal analysis of speckle can be used to estimate the age of fruits after harvest, assuming the correlation function as a sort of parameter to classify differences in speckle.[5]

The autocorrelation function of the time history of a speckle pattern has been proposed as a tool to estimate postharvest age of oranges.[5] It was shown that its full width at half maximum (FWHM) value from samples with different harvest time varied from 0.33 seconds with 1-day-old fruit to 0.97 seconds with 2-week-old fruit.

This variation occurs because in the whole period of observation, the activity of the sample decreases with time, and then the variations of its speckle decrease, resulting in increase of the THSP autocorrelation and, consequently, its FWHM.

Bergkvist[3] presented some different results concerning speckle variation over selected vegetables with a wide variety of characteristics. It was reported that three groups of fruits were gathered with respect to their peels, and after laser illumination, the THSP patterns were analyzed by the FWHM index of the power spectrum and the ratio of Lorentzian to Gaussian width ($Y_{L/D}$). Fruits and vegetables with thick peels were banana, mango, potato, and pumpkin; the group of hard peels was formed by aubergine, avocado, and cherimoya, and the third group with a very thin peel was represented by green apple, lettuce, red apple, and strawberry. It was noticed that fruits and vegetables with thick peels have typically higher values of FWHM and $Y_{L/D}$. When the specular reflection of light is eliminated by turning a polarizer, the values of FWHM decrease. Typical values for hard peel vegetables are small, but for thick peel ones, the FWHM increases to 0.65–0.85 Hz. It may be due to the fact that laser beam penetrates the objects of thick peel deeper than those with hard peel, bringing information from internal variations that can be higher if compared with surface variations.

Until now we have noticed that speckle variation in fruits depends on the material, its age, and its surface characteristics. Is there any other kind of phenomenon that can also cause speckle variation in fruits? In order to answer this question, it is necessary to consider what happens within cells when a fruit, for example, is growing or is getting aged. We know that the cellular activity suffers modification that is reflected on its biospeckle. Therefore, it may be expected that another kind of phenomenon that alters biological activity can be studied by speckle process, such as the study of injury—for instance bruises in biological material like fruits and vegetables.[7] After a mechanical impact, the biospeckle variation on the apple is lower than before, and the variation decreases as the time increases after the impact, even when the human eye is not able to distinguish any variation, for example, on a market shelf. In the reported case, the inertia moment before the impact reached 804.1, and immediately after the impact, the value decreased to 703.0; after 1 hour the values scored 690.7. Finally, after 1 day from the impact, the inertia moment (IM) was 663.1.

It can be observed that it is very simple to apply biospeckle in fruits, although the experimental design should consider the specificities of the samples. Therefore, every kind of fruit should have a specific treatment and design, for instance, taking into account the thickness of the peel, as shown before. In addition, depending on the geometry of the specimen, the intensity of the laser may be difficult to adjust due to the curvature of the surface and the distribution of light over it.

Some specific features relating to the biospeckle phenomena and fruit illumination by laser are presented as follows, with some interesting details on experiments.

8.3.1.1   Points of Illumination

Can we expect differences in the time history of the biospeckle over the surface of biological objects? If we imagine a fruit with uniform distribution and composition, the answer is no. Nevertheless, biological material is very complex, and the distribution of its constituents is not uniform within the whole sample. We can expect that the biospeckle will be different over the illuminated surface, with the differences in color and scars being additional factors to be considered. In order to show these differences, some experiments with oranges were reported.[6]

Four points were chosen over the surface of oranges to test their biospeckle activity. Points A and B were located in the equatorial region, point C in the peduncle insertion, and point D placed in the apex region of the fruit. Figure 8.3 shows the points over the surface of the oranges (respectively, A, B—equatorial region; C—peduncle insertion; D—apex).

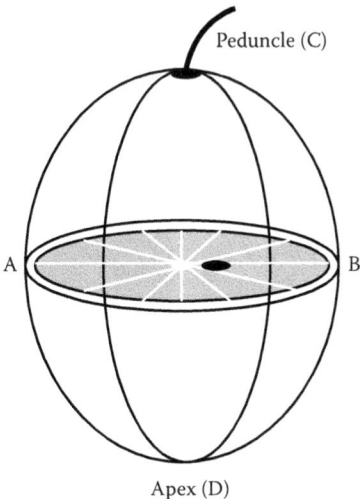

FIGURE 8.3 Points over the surface of oranges (respectively, A, B—equatorial region, C—peduncle insertion, D—apex)

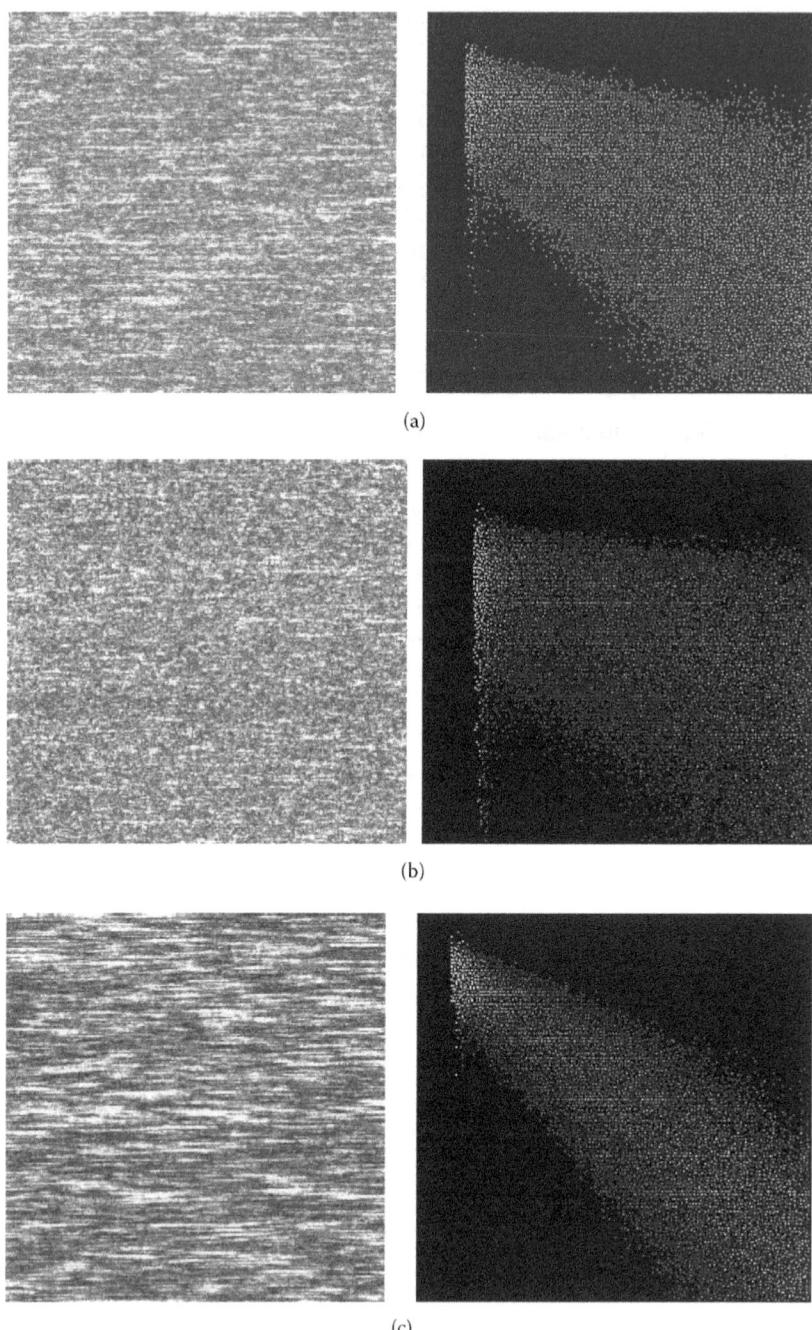

FIGURE 8.4 STS and MCOM for images taken at positions: a, b—equator, c—peduncle insertion, and d—apex (4 days).

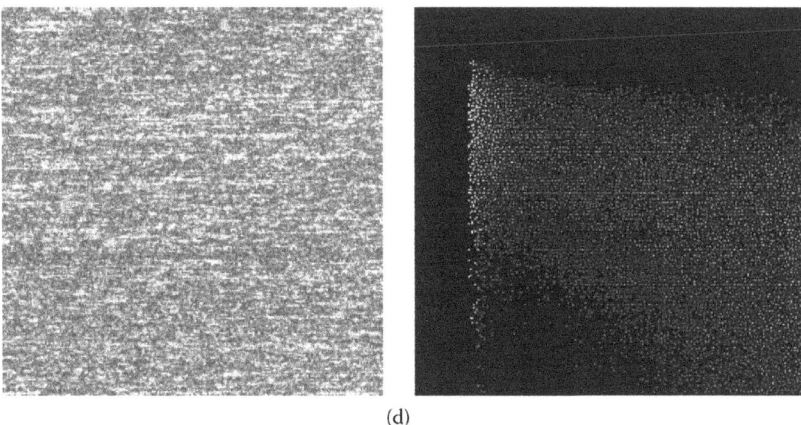

(d)

FIGURE 8.4 (*Continued*)

The results from laser illumination carried out on oranges are represented in Figure 8.4 with the time history speckle pattern of 512×512 pixels. Those diagrams show that the fruit exhibits different THSP patterns, depending on the illuminated point observed by the level of continuity throughout the rows. The inertia moment associated with the images was calculated, and the average values for 24 specimens of four different ages (2 days postharvest, 4 days postharvest, 7 days postharvest, and 14 days postharvest) were assembled.

As it can be observed, point D was the most representative, the apex point. Why is the apex able to better identify differences between ages than the other regions? It is known that the sugar distribution in orange is more concentrated in the apex; as the orange ages, the sugar suffers a degradation process that may be strongly correlated to inertia moment values: 14 days—1861.5; 7 days—2648.0; 4 days—3021.4; and 2 days—3598.0.

Furthermore, the IM values for other points of observation in oranges are different in absolute value and cannot be statistically distinguished. This is the case of the peduncle insertion at point C, where the following values were found: 14 days—940.0; 7 days—1673.1; 4 days—452.9; and 2 days—1250.9.

Why is this point nonsensitive to changes that occur in the orange? The answer could be the constitutive material of point C, which is mainly made of pectin. Thus, point C shows low activity because it behaves as a scar where the peduncle is inserted.

8.3.1.2 Specular Reflection

Many fruits present a surface with polish peels where specular reflection is intense and may undermine the information from the tissue near the surface under monitored. Tests conducted in oranges[6] lead to the conclusion that the specular reflection beam did not affect the results of IM values.

Because the specular reflection does not change the phase of light, the portion of specular reflection in the scattered light can be eliminated from the results if a polarizer is adjusted between the sample and the camera.

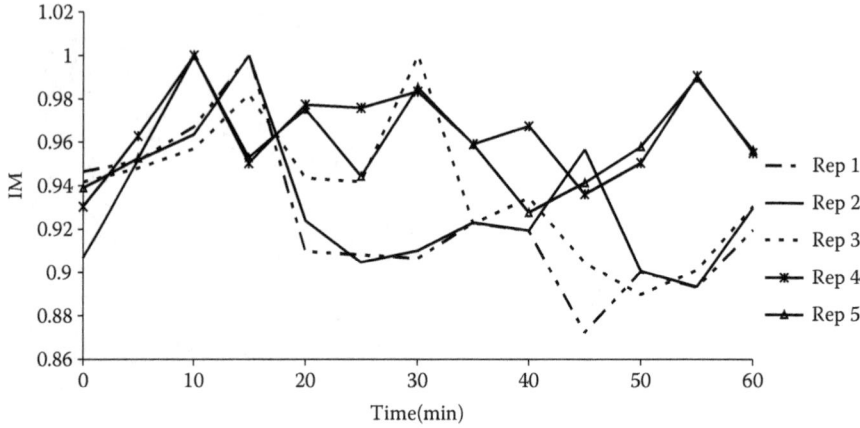

FIGURE 8.5 Regression points for IM versus exposition time variation.

8.3.1.3 Exposition Time to Laser Beam

A frequent question concerning laser illumination upon biological specimens is about the effect of light on the behavior of the sample, which could interfere with the monitored features.

One study[6] evaluated the influence of the period of fruit illumination in speckle patterns, and the results showed no relation between laser and increasing or decreasing of activity, which boosts the use of biospeckle in fruits. This issue should be reconsidered for every other biological specimen under low power laser beam illumination when there are not enough studies to confirm any behavior. Figure 8.5 shows the evaluation of laser influence in tissue activity and its absence of interference with the monitoring of biospeckle phenomena.

8.3.1.4 Daily Variation of Biospeckle in Oranges

The senescence process of fruits depends on the postharvest time, and because the biological activity changes as the postharvest time increases, one can expect that the biospeckle follows the biological variations. The monitoring of such a feature using dynamic laser speckle has been reported in the literature[6] and has shown reliable results, representing a potential tool to be used in fruit packing houses. In the reported experiment, a group of 30 oranges was collected and submitted to a laser beam, and the inertia moment was calculated on certain days. Statistical tests were applied, and the average values of inertia moment showed differences during the first 3 days. After this period there was a tendency for stabilization of values. Considering the conclusions obtained, it is possible that the biospeckle results from the moving of chloroplasts and mineral particles; it can also be assumed that this movement depends on the soluble solids concentration. Orange fruits are nonclimacteric, which means that the maturation process finishes as soon as the fruit is collected from the tree. Thus, the quality of the fruit becomes poor, and the components suffer a

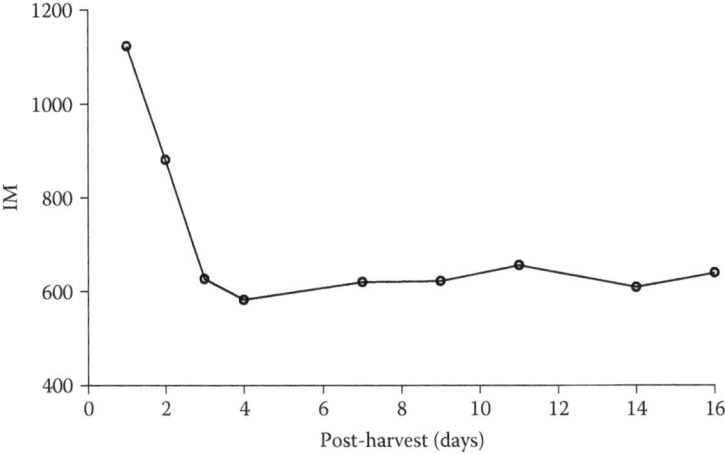

FIGURE 8.6 Regression points for daily variation of inertia moment.

degradation process. This process can interfere with the speckle, but after some days the daily variation can be slow; the equipment may not be capable of detecting these short variations after 3 days. Figure 8.6 shows the regression points for daily variation of inertia moment.

8.3.2 SEEDS

Seeds have a complex and varied constitution. Even in their main constituents we can observe some diverse compositions occurring in the embryo and endosperm structures. The embryo is the portion of the seeds where life is present, waiting for the required conditions to start the germination process, which will consume the reserve of food stored in the endosperm. Some seed, such as corn, show an embryo structure physically separated from the endosperm, which is dead tissue. In other seeds, such as beans, the reserve of food of the endosperm is mixed with the live cells of the embryo. In this complex structure, the analysis of its vigor and viability have presented a difficult challenge to physiologists attempting to create some protocols to evaluate the vigor and the viability condition of seed.[8]

Viability is a quantitative index provided to grade a seed from the point of view of germination, and vigor is an attribute of quality that evaluates functions of the seeds in terms of germination, storability, and behavior in the field; it is a qualitative value.[9]

Despite efforts to obtain a single test as a standard, research has not yet offered an alternative to using distinct approaches. Therefore the scientific community is giving major attention to developing new tests as to improve the reliability of the existing ones. [10–13]

Before reviewing dynamic speckle research in seed analysis, it is interesting to discuss the effect of moisture, an issue that involves the use of laser technology as a tool to evaluate vigor and viability.

8.3.2.1 Moisture in Seeds

Water is a very important variable in seed analysis, and that is also true for many other applications involving biological materials. Therefore, a detailed discussion is useful, which could be easily adapted to other applications, providing a basis to evaluate the dynamic speckle potential as a tool to measure moisture content.

8.3.2.2 Water Inside Seed Tissues

Water is a plentiful element in plant cells. Nevertheless, in contrast to what happens in other plant constituents, in mature seeds the content of water is low. Moisture content in a seed will vary according to its maturation stage, and the water level will affect all the biological processes. A guarantee of good storage can be achieved by keeping the moisture content interval between a 10 to 13% water basis (wb), whereas the intense activity and the attack of insects and microorganisms appear between an inner moisture content of from 18 to 45% (wb).

Water, the best known and most plentiful solvent, is an ideal medium to allow the motion of molecules inside and through cells, which is fundamental to cell and tissue existence.[14–16] One characteristic of the water molecule that is special to its interaction is its polarity, which allows the mutual attraction that forms hydrogen bonding. Such bonds are related to the mobility of the molecules in inner tissue and cells, and to the power levels of the linkages.[14, 16–18]

Water inside seeds has been classified into four groups related to their bondings and number of polar clusters. Physiologists have adopted a classification to link water inside seeds at four levels: constitution water, adsorbed water, solvent water, and absorbed water. Figure 8.7 shows a schematic visualization of these four levels.

Absorbed water level is water inside capillaries and empty spaces with weak link and high mobility. This water can be easily removed from the seed during the first drying stage.

At the level of solvent water many substances dissolve, causing the seed to be susceptible to fungi and various chemical reactions. Therefore, the water needs to be removed during the drying process before storage.

Constitutional water and adsorbed water show covalent bondings, with water reducing its mobility in comparison with the other water levels, which do not permit any assumption of their relation with the majority of the movements inside tissues and cells.

The main hypothesis is that determining solvent water is the aim of the dynamic speckle method of moisture content evaluation. Absorbed water otherwise represents a problem in dynamic speckle evaluation as it produces a high variation on the inertia moment values as a noise that undermines water activity measurement inside the seed. On the other hand, as absorbed water is easily removed, the routine protocol is to start the analysis of seed using dynamic speckle only when that water has been removed.

Some points should be made clear here, and they are related to different types of linkages involving water and seed components. The macro molecules show distinct levels of interactions with water, and in this case the types of interactions permit the conclusion that we can get seeds with the same amount of water, but without the

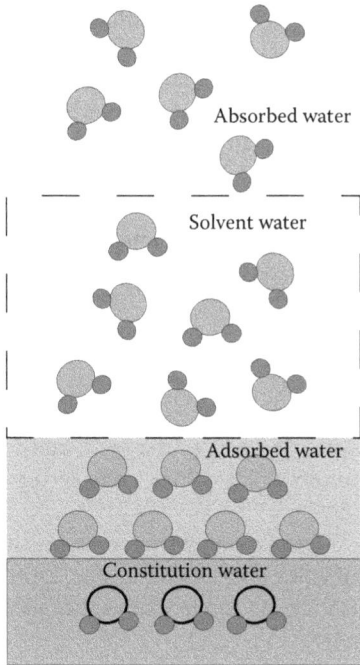

FIGURE 8.7 Schematic distribution of the four levels of water in a seed.

same mobility of the molecules. This is the key to introducing an alternative way to measure water in the seed with respect to its activity.[14–15]

8.3.2.3 Moisture Content versus Water Activity

The literature has already reported the relation between moisture content and dynamic speckle parameters.[19–21] Nevertheless, the use of water activity should be taken into account in new experiments, as the water content also gives information about the motility.

Moisture content can be expressed as a dry basis or wet basis, where the dry basis, $U(db)$, relates the wet weight (w) with the dry matter (d) directly. The index is

$$U(db) = \frac{w - d}{d} \quad \left(\frac{gH_2O}{gDryMatter} \right) \tag{8.7}$$

where g is the gram unit.

Otherwise, wet basis index is the usual form to express the moisture content being

$$U(\%wb) = \frac{100w}{(w + d)} \, (\%) \tag{8.8}$$

Moisture content evaluation is not as simple as might be assumed, because it is necessary to define what type of water will be considered in the evaluation. The starting point should then be made clear, and normally it is considered including the adsorbed water.[15] In addition, not only water can be removed during dry matter calculation, but also other volatiles components interfering with the water amount calculation. Finally, the functional properties of water should also be considered, leading to an index called water activity (*wa*).

Water activity is a thermo dynamical parameter that expresses the water potential free to be involved in chemical and biochemical reactions and fungi growing. It can be expressed by an adimensional magnitude

$$wa = a_w = \frac{p}{p'}$$
(8.9)

where p is the vapor pressure of water in the seed, and p' is the vapor pressure of water at the same temperature and pressure, considering the equilibrium and also water vapor as a perfect gas.

Such an index allows inferences about microorganism development, food stability, and life span of enzymes and vitamins. The ability of water activity to be related to a range of events in seeds and in many other substances (Table 8.1) can be expressed in the literature.[22–24]

8.3.2.4   Overview of Routine Moisture Measurements

Moisture measurement is a key test in seed and grain production and storage, requiring distinct technological solutions with high levels of precision and accuracy, besides robustness and simplicity.

Routine moisture measurements can also be classified in accordance with the transduction process and the ability to give fast measurements. We can point out three methods that are well known and present different characteristics based on all the parameters that we identified. They are the Karl–Fisher, oven, and dielectric methods.

TABLE 8.1
Values of Water Activity to Known Samples

Water Activity (a_w) Range	Substance
1.0	Distilled water
0.97	Milk
Over 0.7	Most fungi
0.6–0.65	Dried fruits
Under 0.6	No microbial proliferation
0.3	Cookies

Source: Fontana, A. J., *Proceedings of the First NSF International Conference on Food Safety*, Albuquerque, 1998, 177; Fennema, O. R., *Food Chemistry*, 2nd ed., Marcell Dekker, New York, 1985, 46.

Karl–Fisher is an example of a chemical transduction method with high accuracy, which is a useful reference to calibration, though highly demanding in terms of human and physical resources. An absence of standardization of its use is also a drawback.[15]

The oven method is known worldwide, nevertheless it does not have its own internationally recognized standard protocol, thus demanding many local standardizations.[25–26] It is a slow procedure that requires at least 16 hours of evaluation with precision limitations.

The dielectric method is another alternative that is popular, fast, and routinary for commercial analysis. Nevertheless, this method has problems related to some ranges of moisture, as well as it presents huge accuracy variations, depending on the incorporated technologies.

These three methods, and many other alternatives to moisture measurement, are available for all uses, though, they still leave room for new alternatives like the dynamic speckle approach, which offers information on the activity of seed water content that is similar to the water activity concept.

8.3.2.5   Moisture by Dynamic Speckle

The ability of dynamic speckle to follow the moisture level in a seed is shown in a visual way in Figure 8.8 with distinct gray levels visible on three cotyledons of bean seeds under different levels of moisture. One seed was illuminated without any contact with water, another in contact with wet paper for 6 hours, and the last one in contact with wet paper for 12 hours. Light gray levels in the image processed represents high activity, whereas the dark gray ones represent low activity. The technique adopted was

FIGURE 8.8 Three seeds under distinct moisture after dynamic speckle image analysis. (From Braga, R.A. et al., *Rev. Eng. Agr.*, 21, 101, 2001. With permission.)

the generalized differences (GD) method (Chapter 5), using 100 images under He–Ne laser illumination with an interval of 0.08 seconds between each frame.[27]

Questions that came up after this technique was developed involved the relationship between the measured activity and the water content or activity. Were the results related only to water or to water plus seed activity, or limited to the seed germination process activity started by water?

These questions are discussed in the following paragraphs, linking the two phenomena, though showing that each one made a significant contribution to the activity. Further questions and interesting comments arise by observing the image, such as: Why is it possible to see gray light in the table? Or did the seed without contact with water present a light grey in one region?

Both questions can be answered by using the same concept, which indicates that the activity in some areas is due to part of the beam that crosses the tissue structures of wet seeds and reaches the table or is reflected on it and reaches another object such as the dry seed. In a formal way, the light scattered inside the seed tissue spreads its changes to other surfaces.

Some information, such as water entering the seed, could be observed after only 6 hours under water contact. It was possible to see the lower activity in the middle of the seed, represented by dark grey, in comparison with light grey in the border of the seed. A bright portion representing the embryo was also noticed, and in this case it was obviously connected with life.

Seed moisture evaluation can be monitored with reliable results by the use of the biospeckle technique as showed in the literature,[19,20,28] which presented results of moisture content measurement of bean seeds under controlled levels of moisture from 11 to 45% wb. The results showed the best efficiency in the range 20 to 45% of water content in the seed, and it was not statistically possible to distinguish the moisture levels under 20% wb. To date, observations indicate that the ability of dynamic speckle is to measure the activity produced by water not linked to any macromolecule but acting as solvent or free in capillaries.

8.3.2.6 Dynamic Speckle and Seed Analysis

Biospeckle laser in seeds can be referenced to in early works presented at the end of the 1990s where the ability of the dynamic speckle to identify activity was first reported.[29–30] The main objective since then, besides the evaluation of other potential applications, has been to use dynamic speckle as a tool to analyze seed vigor and viability. There are many ways to evaluate these conditions. Dynamic speckle can identify or screen distinct areas with damaged tissues from areas in good condition within the same seed. After isolation or identification it is necessary to correlate the location of such damage with its possible influence on the germination process. The limitations presented by routine tests with subjective evaluation, in addition to the hours or days required to implement such routines, were a motivation to do research in this direction.

Earlier studies using biospeckle in seeds tried to conduct the experiments directly to area identification, though the results showed that, at that time, it was only possible to identify separated tissues that were already physically unmatched, such as

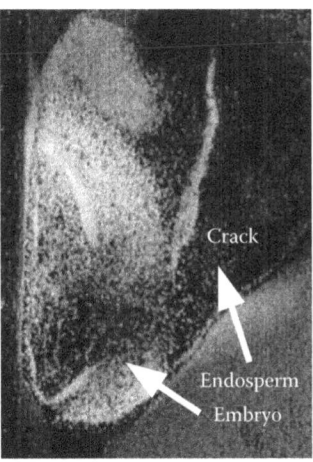

FIGURE 8.9 Corn seed image after GD processing of laser illumination. (From Braga, R.A. et al., *Cien. Agrotec.*, 25, 645, 2001. With permission.)

the embryo and the endosperm in a corn seed, as shown in Figure 8.9.[31] Therefore, identification of damaged tissues inside live and viable structures still presents a challenge, raising various hypotheses and leading to further work.[32]

The experimental designs adopted to illuminate corn seed were based on the back-scattering approach as shown in Figure 8.1, conducted quantitatively and qualitatively. Quantitatively means that IM or autocorrelation were obtained from THSP images. Qualitatively means that the results of the analysis were images, normally produced by GD or Fujii methodologies.

Figure 8.9 was obtained after analysis of 100 frames with a time rate of 0.08 seconds and investigation by the GD method (Chapter 5). The image formed shows in gray levels the activity in each portion of the specimen. The embryo with the highest activity appears in light gray compared to the endosperm without biological activity, which presents a dark gray level.

Besides information on embryo and endosperm, the image shows a clear area inside the endosperm, attributed to water evaporation in the crack observed inside the seed. In addition to water influence, another feature appeared in the results that was hard to explain at first. It was attributed to microorganisms that caused increase of activity in dead tissues, whereas the expectation was just the opposite. Such results forced the use of chemical products to kill the microorganism in order to evaluate the seed without its interference. Similarly, regarding water content, the presence of fungi started to be researched in order to identify and control their influence upon the results. In order to circumvent the interference between damaged and healthy tissue, the decision was made to analyze dead and alive tissues separately.

The success of such a strategy was shown with water loss in dead and live seeds, with activity measured by the IM method.[32–33] A live seed showed higher activity in relation to a dead one, and both reduced their activity with the same behavior during the drying process. These results are expressed in Figure 8.10.

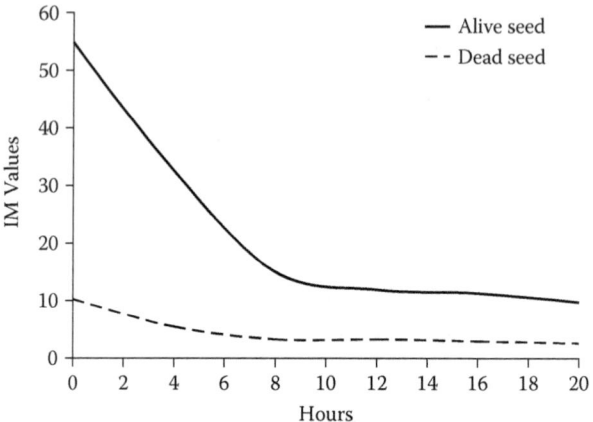

FIGURE 8.10 Curves from beam seeds losing water under laser illumination. (From Enes, A.M. et al., in *Proceedings of XXXIV CONBEA*, SBEA, Canoas, 2005. With permission.)

The answers obtained in the reported experiments allowed some conclusions, such as the reinforced capability to measure water content in seeds and the ability to distinguish dead and live tissues. Along with these answers new questions were raised—for instance, if dead and live seeds present distinct responses to dynamic speckle, why is it not possible yet to distinguish them inside the same seed?

A first hypothesis could relate higher active material undermining lower activity. Nevertheless, there is a report showing the ability to separate dead from live tissues of the same constituency in an apple.[6] So, the shortcomings in seed analysis still remain.

In an inverse way, some reports state that it was possible to detect a higher activity region inside a lower active material. One of them found that an active area (milk) inside a nonbiological material could be observed.[34] Reports concerning the measurement of blood flow in vessels under external observations started to appear the 1970s,[1] showing that, by using dynamic speckle, even in the presence of skin and other stationary or slowly moving scatterers in a tissue, reliable information on high activity inside of low activity material could be obtained (Chapter 5).

Closing the discussion about area mapping, an alternative to circumvent such difficulty could be the use of frequency bands evaluation. Other features should be considered in future investigations on seeds. Some of them are the velocity of the process, the time rate acquirement, and expert analysis.

In this section, the approaches were restricted to the investigation of features in seeds using low power lasers, trying not to disturb the tissues illuminated. On the other hand, the use of a high power laser to intervene, e.g., in seed germination was also observed,[36] and showed that a laser can cause changes in biological tissues.

8.3.2.7 Fungi in Seeds

Forced by the need to control the process of biospeckle in seed analysis, researchers started experimenting with the evaluation of fungi presenting results in both dead and live seeds. This fact turned the evaluation of contamination into one possibility

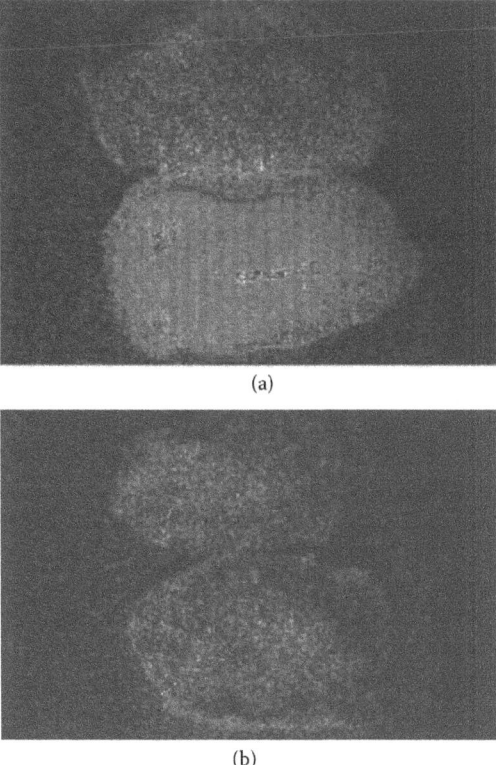

(a)

(b)

FIGURE 8.11 Fujii images of (a) aspergirus over the seed and (b) control seed. (Reprinted from *Biosystems Engineering*, 91, Braga, R. A., et al., Detection of fungi in beans by the laser biospeckle technique, 465–469, Copyright (2005), with permission from Elsevier.)

for the biospeckle application.[28,37-38] Evaluation of biospeckle in dead seeds with known presence of fungi showed the ability to identify two of three types of fungi with respect to seeds without any contamination, which were considered a reference called control seeds.

The techniques adopted were the GD and IM methods, and Figure 8.11 shows the distinction between them using the GD approach. The results obtained by analyzing fungi in live seeds showed the same capability to identify the presence of fungi, and in addition, they could be distinguished in distinct range of IM values. This means that the capability to screen the activity of each fungus with a different level of IM, suggests that research effort in that direction could be profitable leading to potential application in pathology analysis.

8.3.3 ANIMAL REPRODUCTION

In the last few decades, laser light has been extensively used in human and animal medical sciences, including surgery, wound healing, and high-performance laser therapy in various fields of diagnosis and dentistry.[39-41] The potential of laser utilization in the reproductive clinical area was investigated by several groups. In the spermatogenesis process it has been demonstrated that laser can have bio-stimulatory effects upon germ cells, particularly in pachytene spermatocytes and elongated spermatids in the seminiferous tubules, after 28.05 J/cm^2 laser radiations in the testis. Moreover, ultrastructural changes in the Sertoli cell and dissociation of immature spermatids were described while laser irradiation at 46.80 J/cm^2 was being used.[42]

The major experimental advances in basic and applied assisted reproduction were probably due to sperm sex preselection. The technique used is high-speed flow cyto-metric sorting, which is based on measurements of sperm DNA content and involves treating sperm with a fluorescent dye, then detecting by a laser beam (75 to 200 mW) the small DNA difference (3–4%) between X and Y spermatozoa.[43–44]

Recently, a sorting technique for sperm using a laser trap that can separate faster from slower sperm cells was mentioned in the medical research.[45] Motility is regarded as a major criterion related to spermatozoa viability and fertility. It is broadly accepted that faster sperm cells are more likely to successfully fertilize an egg, so techniques that could identify and analyze different motility patterns in the ejaculate should improve the chances of conception via *in vitro* fertilization by ensuring that only the fastest spermatozoa (or better ejaculates) are used.

In the last years some techniques and devices were developed in the field of semen and gamete analysis and manipulation using laser principles. One of these was an apparatus for measuring concentration and motility of light scattering particles,[46] such as spermatozoa in a semen sample, which provides information about different components of the sperm subpopulations velocity distribution. This device comprises a mechanism for producing two intersecting coherent light beams and a mechanism for locating a container for one sample at the point where the beams intersect.

Another system developed was a laser semen mobility and activity analyzer[47] that uses outputs from a photo-receiver, which are converted into pulses for counting and divided to record proportion of mobile spermatozoa multiplied by the velocity of the cells.

A method to measure artificial insemination sperm motility[48] uses a laser to direct light onto a temperature controlled optical cell. The latter contains a microsample of dilute semen. A photomultiplier or detector receives light reflected from the sperm head via an aperture. A programmable timer gates the detector to a high-pass master filter to remove frequency components below one hertz (Hz). The filtered signal is split into three different frequency band paths, which use three elliptic filters. The filter outputs are connected to integrators, which are used to obtain an assessment of the amplitude of spectral components within certain frequency regions. When each integrator reaches saturation level, it is reset to zero, with the reset pulses counted in the respective digital counter and displayed. Information relating to the amplitude of the frequency components above 100 Hz and the full-spectrum signal is derived. The ratio of high frequency HF content to the full spectrum measurement gives a measure of the percentage of motility of sperm in the sample.

The use of laser and optical arrangement to analyze semen is also mentioned in another device[49] that uses a polarized laser of low optical power emitting a coherent beam of monochromatic light directed to an optical system. The system comprises reflecting-refracting components to divide the original coherent beam of light emitted by the laser in two subbeams. The first subbeam has the original reference frequency; the second subbeam provides output diffused light with a different frequency after passing through the sample, and is influenced by the sample characteristics. The beams are joined together to produce a new wave.

In the in vitro fertilization process a system for micromanipulation[50] of the oocyte that uses laser light has been developed. It allows the membrane of an egg cell to be treated with a laser beam having a specific optical wavelength. The laser beam can be used to weaken or break the egg cell membrane, in order to allow penetration of a sperm cell or to assist implantation in the uterus.

Sexing sperm and embryos are powerful tools in animal farm production to enhance genetic selection. A sperm sex selection method[51] is described using thawed frozen semen. The method consists of a flow-citometry sorting of spermatozoa previously stained with a fluorescent dye, and separation of Y and X sperm cells by a laser beam in subpopulations that can be used in artificial insemination or *in vitro* fertilization mainly in the bovine species. The method is effective for the production of embryos from frozen semen and gives the choice of conception to the individual carrying the useful genetic characteristics without time and geographical limitations, because of the use of cryopreserved semen.

Semen is a complex cellular suspension that contains spermatozoa and secretions from the accessory glands of the male reproductive tract. The fluid portion of this suspension is known as seminal plasma. In the human species the first portion of the ejaculate, about 5%, is made up of secretions from the bulbourethral and Littre glands. The second portion derives from the prostate and contributes within the range of 15 to 30% of the ejaculate then follows small contributions of ampulla and epididymis and, finally, the seminal vesicles, which contribute to the remainder, and majority, of the ejaculate.[52] These contributions are similar for several animal species. Apart from providing a liquid vehicle for the sperm cell transport, accessory glands functions are not completely understood because spermatozoa from the pars caudata of the epididymides can fertilize oocytes without the addition of these glands secretions.[52–53] Nevertheless, semen has a very high buffering capacity, much higher than that of most other fluids in the body. Semen maintains its pH near neutral in the acidic vaginal environment, providing the sperm with the opportunity to enter the neutral pH cervical mucus.[52–53]

Some of seminal plasma constituents are citric acid, fructose, sorbitol, ergothioneine, and glycerylphosphorylcholine, and a great variety of hormones, like prostaglandins, androgens, estrogens, antimicrobial substances, and enkephalins.[53]

The rheological properties of human semen change after ejaculation.[51] The initial ejaculate quickly coagulates into a gelatinous substance, and this material then liquefies. Liquefaction occurs over a period of 5 minutes in vivo, but may take 20–30 minutes in vitro, according to some authors. The biochemical mechanisms of this coagulation and liquefaction have been investigated, and it is known that the coagulation factors derive from the seminal vesicles, whereas liquefying factors come from the prostate.

The sperm cells are formed within the seminiferous tubules of the testes and are cells which contain the nucleus, condensed by protamines into a highly compact and hydrodynamic sperm head with a large flagellum to allow its motility. Sperm cell concentration is a parameter evaluated in semen analysis and considered to be related to semen fertility. It ranges from 0.1–$0.2 \times 10^8 \times mL^{-1}$ in the boar, 0.8–$1.2 \times 10^8 \times mL^{-1}$ in the bull, 0.1–$1.5 \times 10^8 \times mL^{-1}$ in the stallion, and 50–$90 \times 10^8 \times mL^{-1}$ in the cock.[53]

Physical analysis of the semen[54–55] consists in evaluation of volume, gross and individual motility, sperm cell concentration and morphology, and presence of unusual elements in the ejaculate such as leucocytes, blood cells, and immature cells of the spermatogenesis lineage.

Basically, in domestic animal species, ejaculation takes place into the vagina or in the uterus during natural coitus. In the former situation, like in the ram, bull, and buck, the semen is highly concentrated and shows strong motility patterns in order to achieve good conception rates, because it is deposited far from the fertilization site (the uterine or Fallopian tubes). In the second case, which occurs in the stallion, for example, the sperm cell concentration is lower and so is the sperm velocity. Spermatozoa shape, size, and velocity vary among species, leading to different physical and motility patterns that are evaluated during semen analysis using light microscopy or computer-assisted semen analysis (CASA).[56–59]

The first approach to verify whether the biospeckle laser could be used as a tool in sperm cells evaluation was carried out using ram (*Ovis aries*) semen and the efforts were concentrated to assess kinetics parameters. The ram sperm cell exhibits head dimensions of 8.2×4.25 µm and 75 to 80 µm of total length.[54] In this species average sperm cell concentration was described[60] ranging from 0.8 to 3.8×10^9 cells $\times mL^{-1}$. In 1948, Lord Rothschild from the Cambridge Department of Zoology first described the activity of the ram semen as follows:

> Active ram semen exhibits a striking phenomenon, sometimes known as "wave formation" which is due to the spontaneous reversible aggregation of spermatozoa in the suspension. These aggregations form and disrupt throughout the semen while the spermatozoa are active. It seemed possible that such macroscopic changes in the "structure" of the sperm suspension might be associated with variations in its electrical properties.[61]

As a matter of fact, this phenomenon is a unique attribute of the semen of ruminants and is thought to be a consequence of high sperm cell concentration and velocity. Presently, it is termed gross motility, *massal motility*, or *swirling motion*.

Thus, the ram ejaculate is characterized by presenting larger cells, compared to other species, high sperm concentration and motility, so it was chosen as the first model to investigate semen evaluation by biospeckle.

Many efforts to develop the biospeckle laser (BSL) technology as a reliable tool to analyze animals' sperm samples have been conducted since the early years of the millennium by a multidisciplinary group at the Federal University of Lavras, Brazil. Some results can be summarized as follows, with the first results showing the positive correlation between the inertia moment, with velocity and gross motility at a level of $p < 0.001$ in ram species (Figure 8.12).

In the ram semen the inertia moment showed correlation coefficients of 0.6552 in relation to gross motility and 0.7210 in relation to spermatozoa velocity. In contrast, higher coefficients ($r = 0.904$) were observed in the bovine frozen semen. Figure 8.13

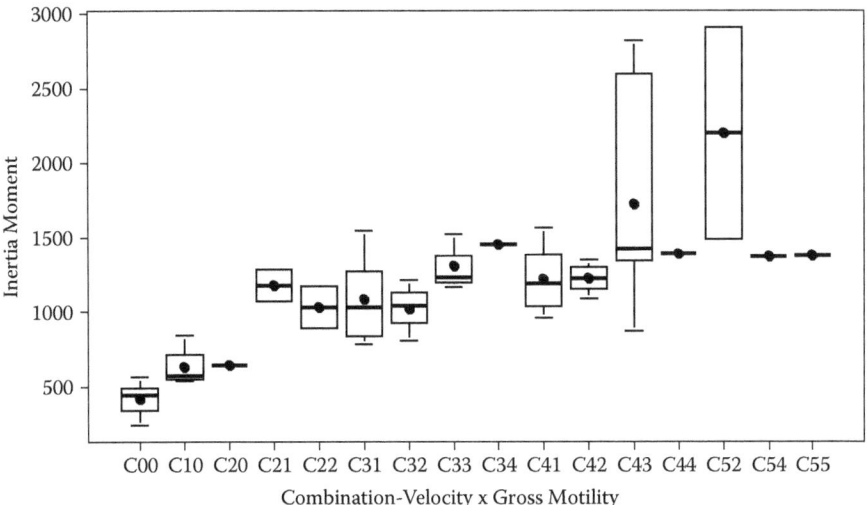

FIGURE 8.12 Box plot for inertia moment data related to the sperm cells velocity and gross motility of the ram semen. C = Combination; numbers following C representing the evaluation of velocity and gross motility in a 0 to 5 scale, e.g., C42, velocity 4; gross motility 2.

shows that when velocity and gross motility values increase, inertia moment also increases, allowing motility sperm cell evaluation by the biospeckle.

The THSP and inertia moment were useful to discriminate different semen motility activities, and when compared to the routine light microscopy analysis they exhibited high coherence (Figure 8.13). In addition, the technique was able to follow the changes that occurred in the sample when the spermatozoa were reducing their activity, caused by a cooling process.

Spatial temporal speckle of ram semen presented in Figure 8.13a shows a sample with an inertia moment of 1170.33, classified as to velocity and motility in 2/2, in a 0 to 5 scale by the light microscopy method. Figures 8.13b and 8.13c have higher inertia moments (1378.32 and 1523.38) and were classified by the same method in 3/3 and 5/5.

Figure 8.13d presents a spatial temporal speckle THSP of low activity ram semen exhibiting an inertia moment of 133. Figures 8.13e and 8.13f show the same ejaculate sample after applying cold shocks in times t_0 and t_1. Changes observed in the STS are due to sperm cells motility progressively decreasing.

Despite the sensitivity of the biospeckle to motility parameters, no statistical correlation was observed ($r = 0.0387$ and $p = 0.9000$) between sperm cell concentration in the ram ejaculate and inertia moment, when 13 ejaculates were analyzed—at least when concentration values ranged from 1.49 to 5.16×10^9 sperm cells $\times$ mL^{-1} (Table 8.2). Nevertheless, further investigations using bovine frozen semen showed that the inertia moment was highly correlated to mobile sperm cell concentration (living cells), and it was not correlated, or at least had a weak influence, to immotile or dead cells present in the ejaculate (total concentration).

FIGURE 8.13 Spatial temporal speckle (STS) of the ram (*Ovis aries*) semen.

FIGURE 8.13 (*Continued*)

TABLE 8.2
Spearman Correlation Coefficients Estimates with Nominal Level of
Significance (p) to Some Physical Characteristics and Inertia Moment of
the Ram (*Ovis aries*) Semen.

	Velocity	Gross Motility	Sperm Cell Concentration	Inertia Moment
Velocity	1,0000	0,8536	−0,0229	0,8812
		($p = 0,0002$)	($p = 0,9408$)	($p < 0,0001$)
Gross motility	0,8536	1,0000	0,2239	0,8329
	($p = 0,0002$)		($p = 0,4620$)	($p = 0,0004$)
Sperm cell	−0,0229	0,2239	1,0000	0,0387
concentration	($p = 0,9408$)	($p = 0,4620$)		($p = 0,9000$)
Inertia moment	0,8812	0,8329	0,0387	1,0000
	($p < 0,0001$)	($p = 0,0004$)	($p = 0,9000$)	

In the bull (*Bos taurus indicus*) frozen semen, it was also observed that the bio-speckle was sensitive to discriminately high and low activity ejaculates and to detect motility decreasing patterns (velocity and percentage of motile cells) throughout time (Figure 8.14), which is very useful to assess semen viability, particularly post-thawing. This is an interesting characteristic because it is generally accepted that semen viability after freezing is related to sperm cell fertility and to the high con-ception rates after artificial insemination. In order to verify sperm longevity after

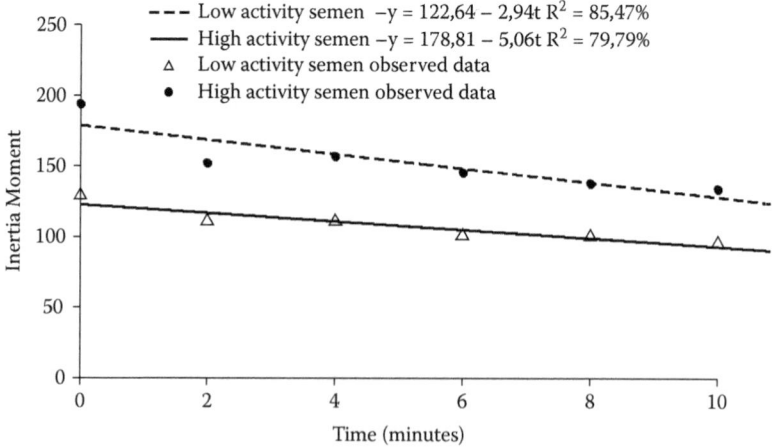

FIGURE 8.14 Mean values to inertia moment in the bull (*Bos taurus indicus*) frozen semen throughout time to high (- - -) and low (—) activity ejaculates.

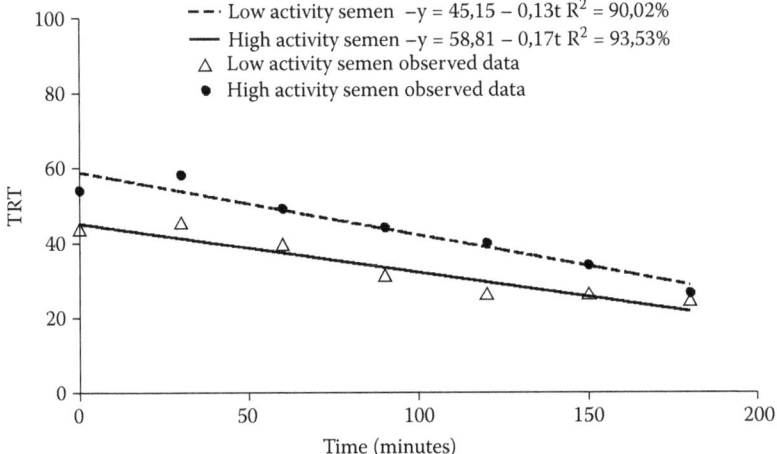

FIGURE 8.15 Thermal resistance test (TRT) mean values of the bull (*Bos taurus indicus*) frozen semen throughout time to high (- - -) and low (—) activity ejaculates evaluated by light microscopy.

freezing and thawing, most laboratories use the thermal resistance test (TRT), which consists of motility evaluation immediately after thawing and 5 hours subsequently (Figure 8.15). The property of the biospeckle to detect decreasing activity in the semen could be also an alternative method to TRT analysis.

Finally, biospeckle is influenced by the experimental setup, the source or wavelength of laser light, spermatozoa morphological differences among species, and by the sperm movement pattern and intensity. In conclusion, it should be considered that biospeckle laser is a sensitive optical approach that could be used in routine and research semen analysis to assess motility sperm parameters.

8.4 PARASITES MONITORING

There are several reasons to study parasite motility. The first ones are related to the evaluation of the action of drugs on such microorganisms, and to test for the presence of parasites in a sample.[62] The reported results in the literature used the forward-scattering geometry in the experimental design.

In order to avoid the saturation produced by the laser beam directed to the CCD a beam expander and a diffuser were placed between the laser and the tray with the sample of parasites.[62] Figure 8.16 shows an apparatus proposed to measure microorganism motility. A limitation of the technique is that low concentrations of parasites do not produce observable activity.

The reduction of the tray size and the study of parasite behavior under laser light should be the next steps to investigate in order to get reliable information from samples with low concentration. The contribution of the diluents to the results should also be considered in further experiments.

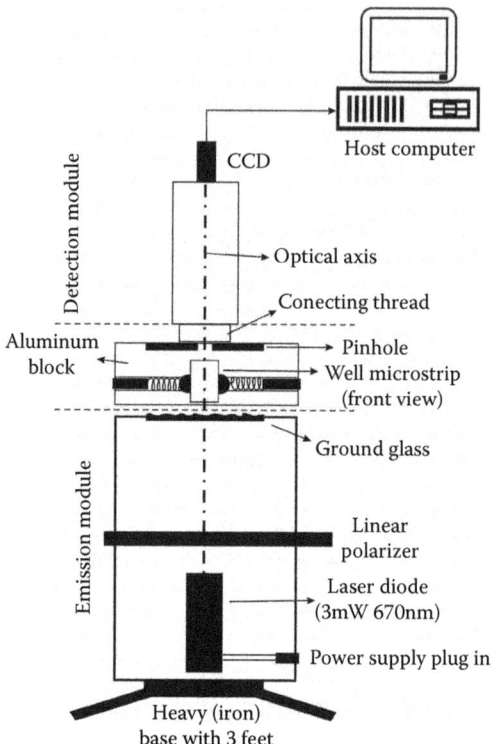

FIGURE 8.16 Forward-scattering arrangement to measure parasites. (Reused with permission from J.A. Pomarico and H.O. DiRocco, *Review of Scientific Instruments*, 75, 4727 (2004). Copyright 2004, American Institute of Physics.)

8.5 MEDICAL APPLICATIONS

Hundreds of references about applications of the dynamic speckle principles to measurements of magnitudes concerning medical applications in the human body can be found in the literature. Early reports were presented in the 1970s,[1] becoming the basis of many efforts to apply the technique known as *laser speckle contrast analysis* (LASCA), in addition to efforts to develop and circumvent the limitation of the spatial limitations of LASCA.

Nevertheless, other approach line using the dynamic speckle information was presented using a methodology which takes the name of the first researcher to present it,[63] the Fujii method, also discussed in Chapter 5. It is observed that in medicine, the dynamic speckle technique has been called by several names, including laser speckle imaging (LSI or LSPI).[64]

Some accounts[1, 65] present the capability to work in real time, or nearly, allowing the observer to follow the changes in the tissues during life activity and to correlate them with other phenomena under evaluation, giving almost instantaneous results. The capability to show the changes online is obtained at the expense of a reduction in the image resolution, which is the main limitation of LASCA. Some studies seek to circumvent this limitation by improving the static image with manipulation of a set of images in time, reducing the real time capability, but with refreshment at a speed that could be considered satisfactory to human perception, and without any loss of spatial information.[66]

The improvement obtained including the time dimension in the analysis, presents a limit that is related to the movement of the human body interfering with the detection of slower changes inside the tissues.

Capillary blood flow in rats or pigs[66–70] as well as in human beings[71–74] has been monitored using speckle activity phenomena. Dynamic speckle has shown to be a useful and reliable tool to measure blood flow inside tissues in spite of the contributions of the activity of the surrounding layers and vessels. One of the accounts[68] showed the ability to measure the cerebral blood flow (CBF) through a (partially translucent) intact rat skull using temporal speckle contrast, or with the thinned preparation of the rat skull[69] in order to evaluate cerebral hemoglobin concentration (HbT) and CBF.

The Doppler effect is the major contributor to activity values registered in capillary blood flow using dynamic speckle, with the other possible changes in the scatterers being concealed by the regular flux. Therefore, the LASCA method, using just one image, takes advantage of a phenomena that is considered best to be monitored by dynamic speckle,[75] despite the complexity of human tissues.

Blood flow measurements were recorded in applications such as retina microcirculation,[65] cadaveric human aortas requiring atherosclerotic plaque evaluation,[73] and the monitoring of fluctuations of light scattered on human erythrocytes in suspension.[71] Other reports of specific usage can be seen in a new instrument proposed using endoscopic and laser imaging to achieve internal tissues,[76] in an application on a knee evaluation using endoscopic laser speckle imaging eLSPI.[63] The usage in skin damages evaluation,[1,77] or skin thermal modification,[78] can be pointed out as additional examples.

One report on the use of speckle imaging proposed the simultaneous usage of photodynamic therapy in tissues to acquire information for evaluating dynamically the changes in blood flow without interrupting the therapeutic approach.[79]

Perfusion phenomenon was pointed out adopting the laser Doppler concept,[77,80–81] in some cases cited as laser Doppler perfusion imaging (LDPI).[82] The perfusion calculation is based on the signal processing of the intensity fluctuation $I(t)$ related to each pixel of the image.[81] The zero and first moment (M_0 and M_1, respectively) of power density spectrum $S(v)$ were calculated using the expression in Equation 8.10 that presents the power spectrum of the signal

$$Power\ spectrum = S(v) = \left| \int_0^\infty I(t)\exp(-i2\pi vt)dt \right|^2 \tag{8.10}$$

The zero-moment related to the average concentration $\langle C \rangle$ of the moving particles in the sample can be obtained by Equation 8.11

$$Concentration = \langle C \rangle \propto M_0 = \int_0^\infty S(v)dv \tag{8.11}$$

And, consequently, the perfusion or flux is obtained by the calculation of the first moment, which is proportional to the speed of the particles in root-mean-square notation (V_{rms}) times the average concentration $\langle C \rangle$, Equation 8.12.

$$Perfusion = \langle C \rangle V_{rms} \propto M_1 = \int_0^\infty vS(v)dv \qquad (8.12)$$

Finally, one report of a patent involving perfusion terminology presented the CMOS image sensor as the instrument adopted to images provenient from laser Doppler perfusion.[83]

8.6 FOOD APPLICATIONS

Dynamic speckle in food analysis has been an interesting tool to evaluate changes in food samples, i.e. during the drying and melting process, as well as to monitor shelf life attributes. By measuring activity by biospeckle, food scientists can evaluate the behavior of distinct blends and their advantages with respect to food stability, for example.

8.6.1 COOKIES

Cookies in general demand some sort of special attributes such as texture and crispness. Such attributes should be maintained for a long time to guarantee a viable shelf life. The relationship between food and environment is critical, determining food stability and ability to offer the same characteristics for long time. Evaluation of the molecular mobility to study food behavior requires many distinct tools, unless one alone is enough to follow all the changes in the structures of the matter and in its bondings.

A study using dynamic speckle as a tool to evaluate the behavior of a special sort of crispy and airy cookie, known as sour cassava starch biscuit, was conducted[84] by establishing a relation between the humidity level inside the storage and the stability of the cookie. Four levels of humidity in a closed ambient were tested, and the results showed the reliability of the tool to follow the evolution of the molecular changes over time as modified by the absorption of water from the air.

Inertia moment (IM) was the method used to evaluate the molecular mobility under biospeckle view, and the report analyzed the samples over 15 days, with 3 replications. It was possible to distinguish the behavior of four solutions (blends), which helped the specialists to choose the best one.

8.6.2 FILMS

Biological film is an important material for food protection and an alternative to plastic in shopping bags, etc. The challenge to research is to monitor the drying process for the optimal blending that is a major factor in its production.

Dynamic speckle has proven to be a possible instrument to follow drying processes as water is a variable that highly influences the dynamic speckle phenomenon, particularly the mobility of the scatterers, in this case, the level of molecular movement.

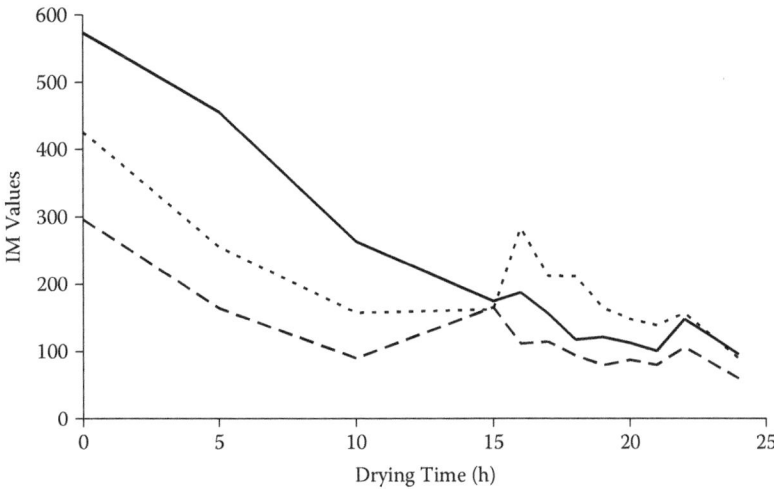

FIGURE 8.17 Drying process of three distinct blends of biofilms under biospeckle observation.

Tests comparing distinct blends and their attributes presented one important application of the biospeckle phenomenon in food analysis, comparing drying time with the composition of the blend.[85] Figure 8.17 presents some advanced results of the drying process observed in three different compounds under dynamic speckle evaluation by IM values. The results show that it is possible to distinguish blend behavior during the drying process as a classification index of drying velocity and stability.

Another interesting potential application of dynamic speckle is the evaluation of inhomogeneities in the whole sample during the drying process, which can be implemented by using generalized differences and Fujii methods to map the areas.

8.6.3 ICE CREAM

Another application to evaluate types of blend optimization using dynamic speckle techniques is the melting of ice cream. Because monitoring the mobility of scatterers is within the skill range of dynamic speckle methodology, the potential exists to monitor the melting velocity of different blends as a reliable alternative.

Emulsions are the base of a wide variety of natural and manufactured materials, including food (milk, creams, mayonnaise, butter, etc.), pharmacological products, biological fluids, agrochemicals, and cosmetics.[86] They are obtained through the mixing of two nonmiscible liquids, and one of them is dispersed into the other in the form of little bubbles. The dispersed phase (also called *uncontinuous* or *inner*) is a liquid in the form of droplets. The surrounding liquid layer is known as the continuous or external phase. Ice cream is a complex colloidal system englobing ice crystals, air bubbles, aggregated (or partially coalesced) fat globules, and a water cryo-concentrated phase. The formation and stabilization of different microstructures involve every ingredient of the formulation. The structure of ice cream determines important parameters for products, such as hardness, resistance to melting, and texture. The structural arrangement is due to different stages of processing, including warm-up, homogenization, pasteurization, maturation, freezing, and hardening, as well as to the components used in the formulation.[87]

Sugars are other elements of great importance. Besides giving a sweet taste, they reduce the freeze temperature of a mixture (the cryoscopic effect) and create a cooling feeling in the mouth during the chewing process. They strongly influence the viscoelastic behavior of the system.

A set of experiments were conducted in which two formulations containing glucose and fructose, respectively, were tested. The replacement of a large amount of sucrose by glucose modifies substantially the viscosity of ice cream, and this fact is directly associated with the nature of the molecules present in emulsion; larger molecules are heavier and increase the viscosity of the mixture. Besides, if these wide chains are not branched, the prospects of side contact between molecules are greater. This intense interaction affects the viscous behavior of the system. As for fructose, the fragmentation of polymeric chains produces a big number of minor molecules and reduces the viscosity.

On the other hand, more molecules interacting with solvent reduces the freezing temperature. Therefore, ice cream manufactured under these conditions melts more quickly when it is taken out of storage at its manufactured temperature, and loss of the confined air in the emulsion occurs. Because of the nature of their molecules, carbohydrates have a good capacity to construct intermolecular hydrogen bridges with proteins. The results of these associations are reflected in the structure of the extruded product. Proteins are amphiphilic groups and have the property of stabilizing emulsions. When the level of adsorbed proteins in the solid-air interface during the manufacturing of ice cream is low, this effect causes an impact on the destabilization of fat inside the formulation, modifying its textural characteristics and consequently its quality.

The use of a nondestructive method that offers a low-cost monitoring transition phase process with this type of emulsion is very desirable because it could be employed in industrial production where quick analysis and operational viability are essential elements. Optical techniques involving dynamic speckles could be a good alternative for meeting this need, as the required equipment is low cost and the assembly complexity is not critical.

Figure 8.18 shows a curve representing the mobility, or activity, of ice cream, identifying some stages that can be useful during the comparison under exposure to different temperatures and among different blends. It is possible to observe that, in the early minutes of monitoring, the inertia moments indicate low activity. This behavior is caused by the predominance of the solid phase, and therefore of less mobility of the scatterer centers. Soon, the ice started to melt and the activity increased, subsequently decreasing. In the next stage, there is a new increase in activity, followed by a decrease to the same level shown in the beginning of process. The second peak of activity was more intense than the first, indicating that the scatterer centers had higher velocities, or that they were larger, which caused larger displacements in the emulsion. This peak is probably related to coalescence and drainage of air bubbles, due to higher flowing of the system with the advance of phase. On the other hand, the formation of air bubbles on the surface of the sample was also observed.

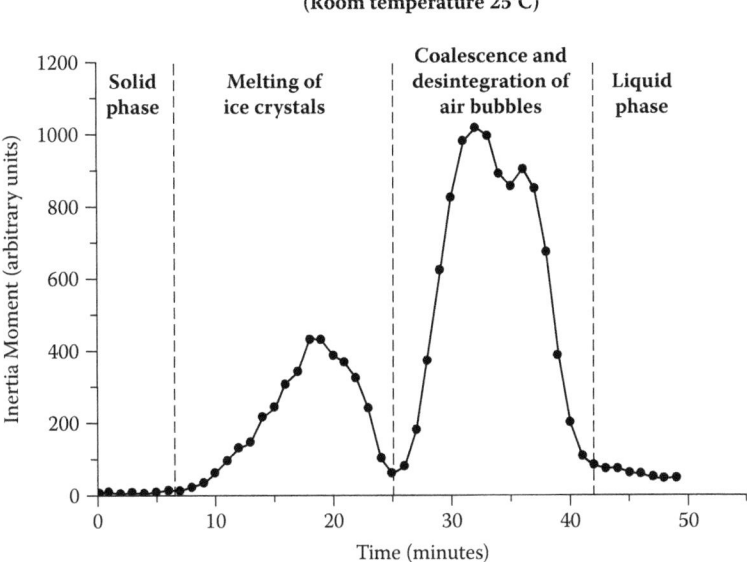

FIGURE 8.18 Stages of ice cream with respect of motility of its components.

Two mixtures of ice cream tested with the dynamic speckle technique showed distinctly different behaviors in their response to changes in time and to three different room temperatures. In Figures 8.19 and 8.20 it is possible to observe such behavior and to compare the changes in the profile of the curves during melting.

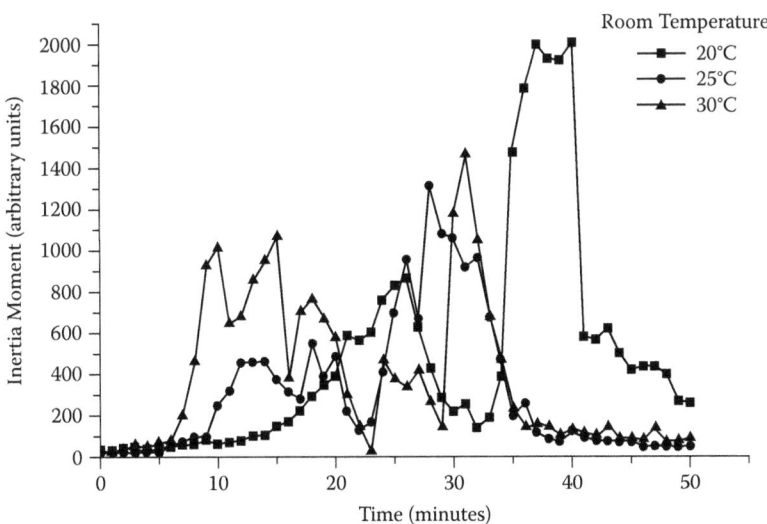

FIGURE 8.19 Blend 1 with changes of biospeckle related to temperature.

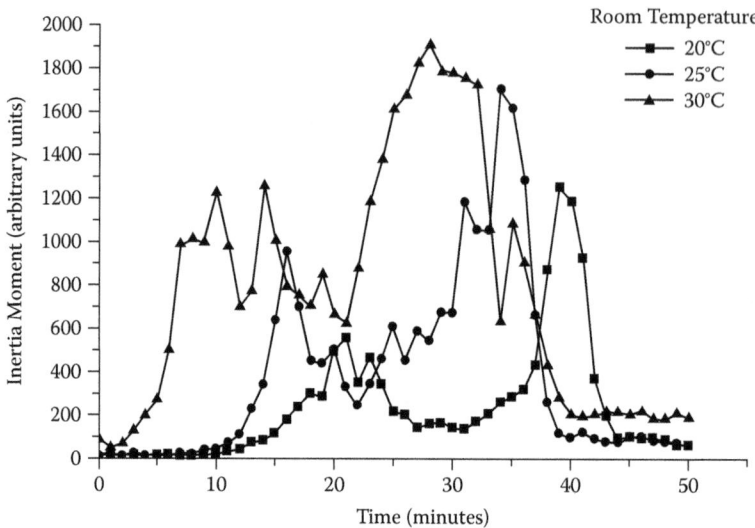

FIGURE 8.20 Blend 2 with changes of biospeckle related to temperature.

The behavior of the samples can be summarized as follows:

- It is possible to observe two peaks of activity.
- The first peak is associated to the melting of ice crystal, and the second one is related to air bubble coalescence and drainage.
- The activity in the second peak is higher when compared to the first, reinforcing the hypothesis that its origin is the coalescence and drainage of bigger structures (air bubbles).
- The distinction between the peaks is better at lower room temperatures.
- The velocity of the process is higher when room temperature increases.
- Activity increases more quickly in the second formulation at all room temperatures.

The ability of the dynamic speckle to follow changes in this sort of material indicates another potential of its use.

8.7 CONCLUSIONS

This chapter aims to summarize for the reader several reports of dynamic speckle in biological matter. Many examples cited here were more completely discussed in previous chapters. Other applications like the references about medicine were only cited, as this area deserves more thorough study in order to collect and explain the many applications presented from a biological point of view. The application in biological areas should involve physicians, physicists, engineers, and other professionals related to the demanded application so that the complexity that links the interaction between coherent light and biological systems can be better understood. All applications demand special approaches and image processing methodologies in their use. Standardization may be pointed out as a need in new developmental areas to advance the design of equipment and further commercial applications.

REFERENCES

1. Briers, J. D., Wavelength dependence of intensity fluctuations in laser speckle patterns form biological specimens, *Opt. Comm.*, 13, 324, 1975.
2. Goodman, J. W., *Speckle Phenomena in Optics*. Roberts & Company, Englewood, 2007.
3. Bergkvist, A., Biospeckle-based study of the line profile of light scattered in strawberries, Master Degree, Faculty of Technology at Lund University, Lund, 1997.
4. Briers, J. D., *Speckle Fluctuations and Biomedical Optics: Implications and Applications, Opt Eng.*, 32, 277, 1993.
5. Xu, Z., Joenathan, C., and Khorana, B. M., Temporal and spatial properties of the time varying speckles of botanical specimens, *Opt. Eng.*, 34, 1487, 1995.
6. Rabelo, G. F. et al., Laser speckle techniques applied to study quality of fruits, *Rev. Bras. Eng. Agr. Amb.*, 9, 570, 2005.
7. Pajuelo, M. et al., Bio-speckle assessment of bruising in fruits, *Opt. Lasers Eng.*, 40, 13, 2003.
8. Marcos Filho, J., Testes de Vigor: Importância e Utilização, in *Vigor de sementes: conceitos e testes*, Krzyzanowski, F. C., Vieira, R. D., and França Neto, J. B., ABRATES, Londrina, 1999.
9. Basu, R. N., Seed Viability, in *Seed Quality: Basic Mechanisms and Agricultural Implications*, Basra, A. S., Ed., Food Products Press, New York, 1994, Chapter 1.
10. Hampton, J. G. et al., Quality seed—From production to evaluation, *Seed Sc. Tech.*, 24, 393, 1996.
11. Neurohr, R. et al., Photon emission—A new method for scanning the quality of food, *Deut. Leb. Rund.*, 87, 78, 1991.
12. Howarth, M. S. and Stanwood, P. C., Tetrazolium staining viability seed test using color image processing, *Trans. ASAE*, 36, 1937, 1993.
13. Carvalho, M. L. M. et al., Pre-harvest stress cracks in maize (*Zea mays* L.) kernels as characterization by visual, X-ray and low temperature scanning electron microscopical analysis: Effect on kernel quality, *Seed Sci. Res.*, 9, 227, 1999.
14. Fennema, O. R. *Química de los alimentos*. Zaragoza: Acribia, 1993. 1095 pp.
15. Multon, J. L., Basics of moisture measurement in grain, in *Uniformity by 2000*, Hill, L. D., Ed., Scherer Communications, Urbana, 2000, Chapter 5.
16. Taiz, L. and Zeiger, E., *Plant Physiology*, The Benjamin, Redwood City, 1991, 565 pp.
17. Campbell, M. K., *Bioquímica.*, Third edition, Artmed, Porto Alegre, 2000. 752 pp.
18. Némethy, G. and Scheraga, H. A., The structure of water and hydrophobic bonding in proteins, *J. Phys. Chem.*, 66, 1773, 1962.
19. Braga, R. A. et al., Seeds characterization by dynamic speckle patterns: A proposal, in *Biological Quality and Precision II SPIE*, Boston, 2000.
20. Rodrigues, S. et al., Efeito da umidade na determinação da atividade biológica de sementes de feijão (*Phaseolus vulgaris* L.) utilizando imagens de speckle dinâmico, *R. Bras. Armaz.*, 30(2), 135, 2005.
21. Braga, R. A., Jr. et al., Assessment of seed viability by laser speckle techniques, *Biosys. Eng.*, 86(3), 287, 2003.
22. Labuza, T. P., Properties of water as related to the keeping quality of foods in *Proceedings of the Third International Congress of Food Science and Technology*. Washington, 1970, 618.

23. Fontana, A. J., Water activity: Why it is important for food safety, in *Proceedings of the First NSF International Conference on food Safety*, Albuquerque, 1998, 177.
24. Fennema, O. R., *Food Chemistry*, second edition, Marcell Dekker, New York, 1985, 46.
25. ISTA. International Seed Testing Association. *International Rules for Seed Testing.* Iguaçu Falls, 2008.
26. BRASIL. Ministério da Agricultura e da Reforma Agrária. *Regras para análise de sementes.* Brasília: SNDA/DNDV/CLAV, 1992. 365 pp.
27. Braga, R. A. et al., Avaliação da influência da umidade de sementes de feijão (*Phaseolus vulgaris* L.) na análise pela técnica do bio-speckle laser, *Eng. Agríc.*, 21(1), 101, 2001.
28. Braga, R. A. et al. Uso do bio-speckle laser como quantificador de atividade biológica—medidor de umidade, identificador de fungos em sementes e quantificador de atividade em sêmen animal ou humano, BR. Patent PI,0,301,926,8, 2003.
29. Rabal, H. J. et al., Laser na Agricultura, in *Energia, Automação e Instrumentação, XXVII CONBEA*, SBEA, Poços de Caldas, 1998.
30. Arizaga, R., Trivi, M. R. and Rabal, H. J., Speckle time evoluation characterization by co-ocorrence matrix analsis, *Opt. Laser Tech.*, 4(3), 1, 1999.
31. Braga, R. A. et al., Potencial do bio-speckle laser para avaliação da viabilidade de sementes, *Ciên. Agrotec.*, 25(3), 645, 2001.
32. Braga, R. A., Bio-speckle: uma contribuição para o desenvolvimento de uma tecnologia aplicada à análise de sementes, Dr. thesis, Universidade Estadual de Campinas, Campinas, 2000.
33. Enes, A. M. et al., Biospeckle laser em tecidos vivos e tecidos mortos de sementes de feijão (Phaseolus vulgaris L.) durante a perda de água, in *XXXIV Congresso Brasileiro de Engenharia Agrícola*, SBEA,Canoas, 2005.
34. Gonik, M. M., Mishin, A. B., and Zimnyakov, D. A., Visualization of blood microcircultion parameters in human tissues by time-integrated dynamic speckles analysis, *Ann. New York Acad. Sc.*, 972, 325, 2002.
35. Sendra, G. H. et al., Decomposition of biospeckle images in temporary spectral bands, *Opt. Letters* 30 (13), 1641, 2005.
36. Watanabe, Y., Wakisaka, Y., and Hitomi, H., Method for improving germination of hard seed by laser beam irradiation and germination improved seed, E.P. Patent 1,568,264,A1, 2005.
37. Braga, R. A. et al., Detection of fungi in beans by the laser biospeckle technique, *Biosys. Eng.*, 91, 465, 2005.
38. Enes, A. M. et al., Frequency analysis of the biospeckle laser, in *6th Ibero-American Conference on Optics*, RIAO Campinas 2007, 152.
39. Reddy, G. K., Photobiological basis and clinical role of low-intensity lasers in biology and medicine, *Lasers Surg. Med.*, 22(2), 141, 2004.
40. Stergioulas, A., Low-level laser treatment can reduce edema in second degree ankle sprains, *Lasers Surg. Med.*, 22(2), 125, 2004.
41. Gouw-Soares, S. et al., Comparative study of dentine permeability after apicectomy and surface treatment with 9.6 microm TEA CO2 and Er:YAG laser irradiation, *Lasers Surg. Med.*, 22(2), 129, 2004.
42. Taha, M. F. and Valojerdi, M. R., Quantitative and qualitative changes of the seminiferous epithelium induced by Ga. Al. As. (830 nm) laser radiation, *Lasers Surg. Med.*, 34(4), 352, 2004.
43. Johnson, L. A. and Welch, G. R., Sex preselection: high—Speed flow cytometric sorting of X and Y sperm for maximum efficiency, *Theriogen.*, 52(8), 1323, 1999.
44. Welch, G. R. and Johnson, L. A., Sex preselection: Laboratory validation of the sperm sex ratio of flow sorted X- and Y-sperm by sort reanalysis for DNA, *Theriogen.*, 52(8), 1343, 1999.

45. Nascimento, J. M. et al., Analysis of sperm motility using optical tweezers, *J. Biomed. Opt.*, 11(5), 44001, 2006.
46. Urban, C. and Seitz, P., Measuring apparatus for measuring concentration and motility of light scattering particles, has cross correlator for receiving signals from photodetectors and processing mechanism for deriving components of velocity distribution, Patent EP1464966-A1, 2004.
47. Eskov, A. P., Arefev, I. M. and Gurilev, O. M., Laser semen mobility and activity analyser—Has output from photoreceiver converted into pulses for count divided to record proportion of mobile spermatozoids multiplied by velocity, Patent SU1154619-A, 1985.
48. Woolhouse, J. K. and Woolford, M. W., Artificial insemination sperm motility measurement method—Carrying out frequency-amplitude analysis in time domain using filters and integrating filtered output, Patents EP95386-A; AU8314977-A; CA1201301-A; US4601578-A; EP95386-B; DE3374083-G,1982.
49. Cirillo, F. and Scotti, C., Seminal liquid spermatozoa mobility and concentration evaluating devices—Uses laser light and optical arrangement with reflecting and/or refracting components to split beam, Patent EP69092-A; IT1171329-B, 1981.
50. Fuhrberg, P. and Feichtinger, W., Micro-manipulation system for in vitro fertilisation—and uses laser beam for local treatment of egg cell pellicule. Patents EP539660-A; EP539660-A2; EP539660-A3; EP539660-B1; DE59208545-G, 1992.
51. Galli, A. et al., Producing high-genetic value embryos of specified sex, comprises thawing frozen solution of genetically-selected spermatozoa, sorting them into subpopulations, and fertilizing oocytes with subpopulation in vitro, Patent WO200151612-A; EP1255812-A; WO200151612-A1; FR2806441-A3; AU200130176-A; EP1255812-A1; KR2002079850-A; IT1317724-B, 2000.
52. Owen, D. H. and Katz, D. F., A review of the physical and chemical properties of human semen and the formulation of a semen simulant, *J. Andr.*, 26(4), 459, 2005.
53. HAFEZ, ESE. *Reproduction in Farm Animals*, sixth edition, Lea and Febiger, Philadelphia, 1993.
54. World Health Organization (WHO), *Laboratory Manual for the Examination of Human Semen and Sperm-Cervical Mucus Interaction,* fourth edition, Cambridge University Press, New York, 1999.
55. Henry, M. R. J. M. and Neves, J. P., *Manual para Exame Andrológico e Avaliação de Sêmen Animal*, second edition, Colégio Brasileiro de Reprodução Animal; Belo Horizonte, 1998.
56. Gravance, C. G, Champion, Z. J., and Casey, P. J., Computer-assisted sperm head morphometry analysis (ASMA) of cryopreserved ram spermatozoa. *Theriogen.*, 49, 1219, 1998.
57. Gravance, C. G. et al., Replicate and technician variation associated with computer aided bull sperm head morphometry analysis (ASMA), *Inter. Jour. Andr.*, 22, 77, 1999.
58. Gravance, C. G., Lewis, K. M., and Casey, P. J., Computer automated sperm head morphometry analysis (ASMA) of goat spermatozoa, *Theriogen.*, 44, 989, 1995.
59. Rijsselaere, T. et al., Automated sperm morphometry and morphology analysis of canine semen by the Hamilton-Thorne analyzer, *Theriogen.*, 62, 1292, 2004.
60. MIES FILHO, A. *Reprodução dos Animais e Inseminação Artificial*, fifthe edition. Porto Alegre, 1983.
61. Rothschild, L., The activity of ram spermatozoa, *J. Exp. Bio.*, 25(3), 219, 1948.
62. Pomarico, J. A. and DiRocco, H. O., Compact device for assessment of microorganism motility, *Rev. Sci. Instr.*, 75, 4727, 2004.
63. Fujii, H. et al. Evaluation of blood flow by laser speckle image sensing, *Appl. Opt.*, 26, 5321, 1987.

64. Bray, R. C., et al. Endoscopic laser speckle imaging of tissue blood flow: Aplication in the human knee, *J. Orthop. Res.*, 24, 1650, 2006.
65. Konishi, N. and Fujii, H., Real-time visualisation of retinal microcirculation by laser flowgraphy, *Opt. Eng.*, 34(3), 753, 1995.
66. Li, N. et al., Cortical vascular blood flow pattern by laser speckle imaging, in *27th Annual International Conference of the Engineering in Medicine and Biology Society*, Hopkins, J., Ed., Baltimore, 2005, 3328.
67. Winchester, L. W. and Chou, N. Y., Blood velocity measurements using laser speckle imaging, in *26th Annual International Conference*, IEEE EMBS, San Francisco, 2004, 1252.
68. Li, P. et al., Imaging cerebral blood flow through the intact rat skull with temporal laser speckle imaging, *Opt. Let.*, 31(12), 1824, 2006.
69. Dunn, A. K. et al., Simultaneous imaging of total cerebral hemoblobin concentration, oxygenation, and blood flow during functional activation, *Opt. Let.*, 28(1), 28, 2003.
70. Cheng, H. et al. Efficient characterization of regional mesenteric blood flow by use of laser speckle imaging, *Appl. Opt.*, 42(28), 5759, 2003.
71. Pop, C. V. L., Vamos, C., and Turcu, I., Fluctuations of light scattered on human erythrocytes—A statistical analysis, *Rom. Jour. Phys.*, 50(9), 1207, 2005.
72. Fujii, H. et al., T., Blood flow observed by time-varying laser speckle, *Opt. Let.*, 10(3), 104, 1985.
73. Tearney, G. J. and Bouma, B. E., Atherosclerotic plaque characterization by spatial and temporal speckle pattern analysis, *Opt. Let.*, 27(7), 533, 2002.
74. Aizu, Y. and Asakura, T., Bio-speckle phenomena and their application to the evaluation of blood flow, *Opt. Laser Tech.*, 23, 205, 1991.
75. Zhao, Y. et al., Point-wise and whole-field laser speckle intensity fluctuation measurements applied to botanical specimens, *Opt. Lasers Eng.*, 28, 443, 1997.
76. Forrester, K. R. et al., Endoscopic laser imaging of tissue perfusion: New instrumentation and technique, *Lasers Surg. Med.*, 33, 151, 2003.
77. Stewart, C. J. et al., A comparison of two laser-based methods for determining of burn scar perfusion: Laser doppler versus laser speckle imaging, *Burns*, 31, 744, 2005.
78. Zimnyakov, D. A. et al. Monitoring of tissue thermal modification with a bundle-based full-field speckle analyzer, *Appl. Opt.*, 45, 4480, 2006.
79. Kruijt, B. et al., Laser speckle imaging of dynamic changes in flow during photodynamic therapy, *Lasers Med. Sc.*, 21, 208, 2006.
80. Serov, A. and Lasser, T., High-speed laser Doppler perfusion imaging using an integrating CMOS image sensor, *Opt. Exp.*, 13(17), 6416, 2005.
81. Wardell, K., Jakobsson, A., and Nilsson, G. E., Laser Doppler perfusion imaging by dynamic light scattering, *IEEE Trans. Biomed. Eng.*, 40(4), 1993.
82. Rajan, V. et al., Speckles in laser Doppler perfusion imaging, *Opt. Let.*, 31(4), 468, 2006.
83. Serov, A. N. et al., Laser Doppler perfusion imaging using a CMOS image sensor EP 1332718, 2003.
84. Ribeiro, K. M., Efeito da composição nas isotermas de sorção e características do biscoito de polvilho. MS degree, Universidade Federal de Lavras, Lavras, 2006.
85. Silva, W. A., Elaboração e caracterização de biofilmes obtidos de diferentes fontes de amigo. MS degree, Universidade Federal de Lavras, Lavras, 2005.
86. Novales, B., et al., Characterization of emulsions and suspensions by video image analysis, *Colloids Surf. A Physicochem. Eng. Asp.*, 221, 81, 2003.
87. Granger, C. et al., Influence of formulation on the structural network in ice cream, *Int. Dairy J.*, 15, 255, 2005.

9 Applications in Nonbiological Materials

Hector Jorge Rabal, Roberto A Braga Jr.,
and Marcelo Trivi

CONTENTS

9.1 INTRODUCTION

After the successful design and construction of the first laser sources, the nice properties of this coherent light began to be used in several new applications, though initially only in scientific laboratories. Among the techniques developed for nondestructive testing, holography and speckle were used to monitor changes in industrial materials and their surfaces, such as deformations, displacements, cracks, vibration, expansion, constriction, oxidation, the drying process, evaporation, loss of mass, etc.

Similar to the evolution of digital speckle pattern interferometry (DSPI), the first historical experiments involved the photographic register in holographic plates. To make the measurements, the intensity distribution in a speckle diagram was stored in the plate, and careful repositioning of the developed plate and measurement of the correlation were done by mean of electronic devices between the stored initial state and the current state in real time.

Because electronic means of detection, storage, and processing evolved, the setup and processing operations became more versatile and easier to implement, and new phenomena and applications became accessible.

Related to dynamic speckle techniques, we have seen in previous chapters certain typical applications in biologic materials. Nevertheless, the involved phenomena are also present in nonbiological examples, including some industrial processes. The contactless, nondestructive and, mostly, nonperturbative testing with real time or almost real time operation are desirable features of the optical approach offered by the dynamic laser speckle to some problems of industrial instrumentation.

Applications in nonbiological matters normally involve materials with low transparency such as inks, metals under corrosion, cement hydration, foams, and mostly opaque materials, creating a natural barrier to the use of the forward-scattering geometry. These materials are also usually more tolerant to the damage produced by radiation used in the process of measuring.

This chapter presents a set of miscellaneous applications using some already described methodologies discussed in the previous chapters. Further discussions about the phenomenon itself and the difficulties encountered are also covered, showing to the reader some of the advantages and drawbacks of the methodology. This chapter also describes a commercial instrument with special use in paint and ink drying, with an interesting set of practical solutions and layout.

9.2 PAINT DRYING

Paint is basically a liquid that, after application to a substrate in a thin layer, is converted into an opaque solid film. It is basically composed of a vehicle or diluents (solvent or water), pigments, binder, and other additive substances to obtain different properties. The vehicle or solvent is a volatile liquid that acts as a carrier for the other components and does not participate in the final state of the film; for instance, waterborne paints use water as vehicle.

Otherwise, solvent-borne paints, sometimes called oil-based paints, can have various combinations of solvents as the vehicle, such as xylene, toluene, and mineral spirit, among others. Some solvents may not be adequate because of environmental and economic reasons.

Another constituent of the mixture that composes paint is the pigment, which is a granular solid that is added to the paint to attribute color and toughness. In order to warranty adhesion of the components and also to confer attributes such as gloss, durability, flexibility, and robustness, the binder, or resin, is included in the paint composition, and could be acrylics, polyurethanes, polyesters, melamine resins, epoxy, or oils.

The drying process is related to the loss of solvent or water by evaporation. It should be observed that, if the same sort of solvent, for example, is added to the paint dried, the solid binders will dissolve again. That reversible action does not happen with water-based diluents. The paint that uses water as diluent is known as latex paint, though the term latex does not mean latex rubber here. The drying process of that type of paint is called *coalescence,* meaning that the solvent evaporates and the binder particles fuse between them.

Many industrial processes involve the coating of substrates with thin layers of paint to impart to it desirable properties such as gloss, adhesive properties, ink acceptability, light sensitivity, and magnetic properties as barriers to prevent corrosion, modify light reflection or heat radiation of a surface, etc.

Drying is the initial part of the coating process. The knowledge of the drying process has practical applications, for example, to determine the shortest time necessary for drying the first layer before the application of the second one. The velocity of the drying process will be related to environment conditions such as air velocity and temperature, or linked to an artificial source of heat or air flux. The drying process behavior is also governed by the dilution level, the size of the layer of paint, and the geometry of the surface, as well as by its type, which also plays a special role in the drying process.

Although several experimental techniques have been applied in the drying of coatings process, gravimetric analysis, related to the measure of weight loss, is the most direct standard method used. However, dynamic speckle techniques have proved to be useful tools for the characterization of the processes in the drying of paint.[1] It is possible to follow the drying of both solvent-borne and waterborne coatings to obtain information on the process, and the results are successfully compared with those obtained from conventional gravimetric measurements.

Dynamic speckle methods have been applied in the drying process both because of their intrinsic interest and because dynamic speckle is a relatively well known phenomenon that has a rather predictable behavior when used with biological materials. Therefore, it can be used as a case study for testing different dynamic speckle methodologies as in Chapter 5. It does not have a large amount of variables and components in the same sample, and there is less variability as is common in biological tissues.

Inertia moment was the first dynamic speckle tool shown to be able to follow the evolution of the mass loss.[1] A high correlation was found between the optical and the gravimetric measurements ($r^2 = 0.9923$).

Wavelet-based related measurements,[2,3] fuzzy-based methodologies, Hurst coefficient,[4] intensity and phase-only correlation,[5] the difference histogram method,[6] empirical mode decomposition,[7] multifractal analysis,[8] spatial phase variance,[9] and the effect of data size in the results[10] used the paint-drying process as a basic experiment to test their reliability as tools to analyze dynamic speckle behavior. Some of these measurements were treated in Chapters 4, 6, and 7.

Chapter 5 also presented an experiment based on the drying of paint over a coin, which was used throughout the chapter as a case study for testing different activity image algorithms. The visual results permitted the observation of the different stages of drying due to the different thickness of the paint layer accumulated over the coin's profile, shown the high sensitivity of the evaluation techniques based on changes in the speckle patterns activity.

Amalvy[11] found that it is possible, by using the inertia moment, to follow the drying of both solvent-borne and waterborne coatings to obtain information on the process as well. The results correlate favorably with the weight loss obtained with conventional gravimetric measurements when the results are conveniently normalized. Because of the nature of speckle experiments, the optical signal is sensitive to phenomena occurring mainly near the top surface.

We will follow the path developed by Amalvy, which presented a schematic description of the drying process and the corresponding dynamic speckle measurements.

Drying is controlled by the diffusion of the solvent through the paint and the evaporation of the solvent from the surface. The initial stage of this process is called the constant slope falling period. The coating behaves as if it were a pool of solvent, and the evaporation rate is determined by external factors. In waterborne coatings, most of the drying takes place in the constant rate period, with the particles adhering to one another by the capillary force of the water bridging between them.

The next stage is a falling rate period in which the drying rate decreases to zero. Its start depends on the film thickness and solvent evaporation. At first the solvent evaporates from a boundary that recedes into the coating as it dries, leaving a porous outer layer behind it.

Figure 9.1 shows a schematic picture of three stages of drying of a latex paint. In the initial state, the laser beam detects only water flux. There is enough water to

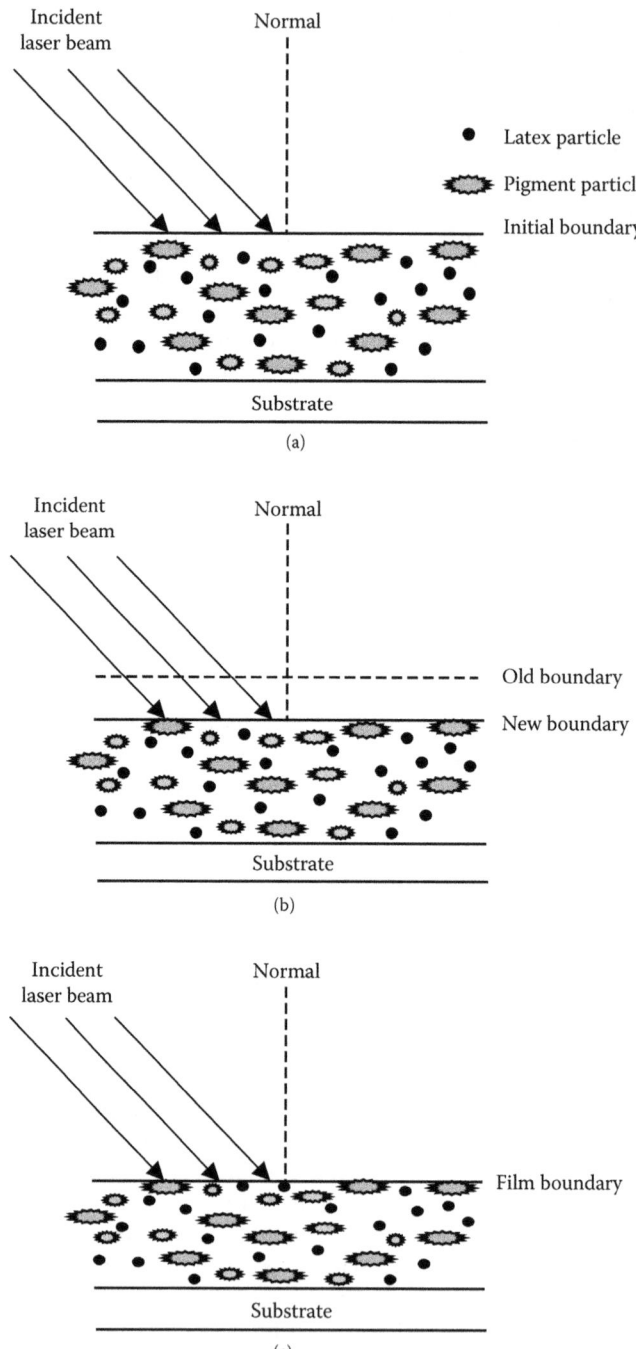

FIGURE 9.1 Schematic picture of the three stages of drying of a latex paint: (a) initial stage, (b) intermediate stage, (c) final stage.

maintain separation of latex and pigment particles. In the intermediate stage, after the surface water evaporates, new material with different dynamic behavior and refraction index begins to appear at the surface. In the final stage, the final packing is almost defined, and only water flux by diffusion is detected.

Figure 9.1 shows that, after the absence of diluent at the surface, the new stage of evaporation is governed by the ability of the diluent to reach the surface by diffusion, capillary flow, or percolation.

The time evolution of the drying of paint was studied using both the normalized inertia moment of the dynamic speckle patterns and the normalized weight loss measured with an analytical balance (±0.1 mg). The results, plotted against the time elapsed since the deposition of the paint, are shown in Figure 9.2. To compare the results, both dependent variables were normalized by scaling them to the 0–1 interval.[11]

In the second stage of the drying process, particles come into contact with each other, forming a closely packed array with water-filled interstices. Loss of interstitial water coincides with particle deformation and compaction. As the latex makes this transition, particles (polymer latex and pigments) begin to emerge at the surface. This has the effect of raising the material density at the surface, and increases the refractive index.

The sensitivity of the speckle method to refractive index changes also enables us to determine the start of different stages of the drying process of latex paints, and therefore, it gives two kinds of informationóone related to the rate of drying and the other related to the change in the materialís components having different refractive indices. This was the case for the waterborne paint where pigments and latex particles have different refractive indices than the solvent.

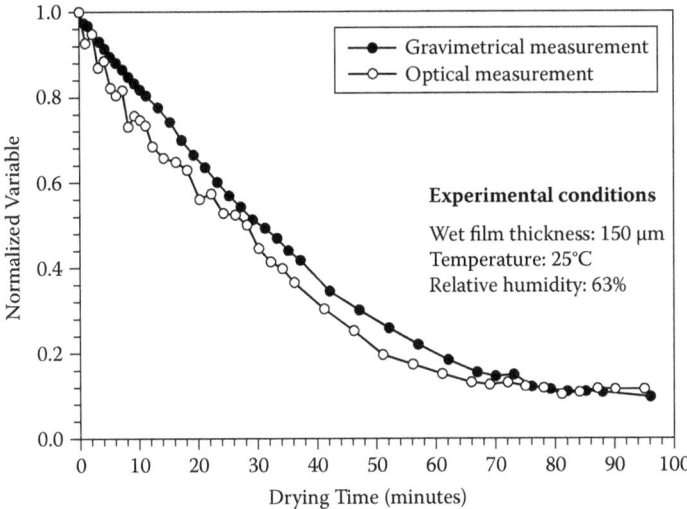

FIGURE 9.2 Curves of white latex paint drying. (Reprinted from Amalvy et al., Application of dynamic speckle interferometry to the drying of coatings, *Progr. Org. Coat.*, 42(1), 89, 2001. With permission from Elsevier.)

Once the application of the dynamic speckle to the monitoring of the drying of coatings is already well established, a commercial instrument is already available. It will be described in some detail in the following section.

The dynamic speckle technique has also been successfully applied to the monitoring of the drying or hardening of other polymers.

9.3 A COMMERCIAL INSTRUMENT (HORUS®)

The fact that a commercial instrument has been developed and offered for specific applications indicates that the underlying technology has reached a certain degree of evolution and confidence. It closes a cycle of research and development.

The commercial instrument, named Horus,[12] uses dynamic speckle to evaluate the paint- and ink-drying process in real time. Its well-developed design uses a CCD camera to acquire the free propagation speckle pattern images both for transmission and backscattering geometries. The data are analyzed using as measurement an index based on the Euclidean distance d between corresponding intensities in every pixel of a pair of image.

$$d = \sqrt{\sum_{x=0}^{dim\,x-1} \sum_{y=0}^{dim\,y-1} \left[I_2(x,y) - I_1(x,y) \right]^2} \qquad (9.1)$$

where $dim\,x$ and $dim\,y$ are the dimensions of each image, and I is the intensity of each pixel at the coordinates x and y. It can be proved that d has a maximum value that corresponds to fully decorrelated speckle images. A decorrelation time τ is defined as the time that would be required to achieve the maximum if the process would act with its initial slope. The reciprocal of this time, τ, in hertz, is called the *speckle rate* (SR). The expression in Equation 9.1 is used instead of the usual correlation function because the former can be performed much faster.

This instrument uses an activity measurement based on a magnitude defined in frequency units. It is an interesting alternative that has also been used in another method.[13] Fast Fourier Transform and Power Spectrum Density are the tools employed to attain the frequency domain.

The device includes, in the same structure, a table used to hold the sample that is also the basement of the device, a mast with a right-angle arm that is used to support the CCD and the laser over the sample at an adjustable distance. It employs free propagation geometry, and the observed speckle pattern is detected by the camera though a narrow interference filter.

The principles used in this commercial device could be implemented also in a set of new applications that could use the same physical structure as well as the same or similar procedure of image processing.

9.4 CORROSION

Corrosion is deterioration of essential properties in a material due to reactions with its surroundings. In the most common use of the word, this means a loss of an electron of metals reacting with water and oxygen. Weakening of iron due to oxidation of the iron atoms is a well-known example of electrochemical corrosion. This is commonly known as rust. This type of damage usually affects metallic materials, and typically produces oxides and salts or both, of the original metal. Corrosion also includes the dissolution of ceramic materials and can refer to discoloration and weakening of polymers by the sunís ultraviolet light.

Most structural alloys corrode merely from exposure to moisture in the air, but the process can be strongly affected by exposure to certain substances. Corrosion can be concentrated locally to form a pit or crack, or it can extend across a wide area to produce general deterioration. The corrosion process is basically due to changes in the surface, as also is the special case of the electroerosion. Changes in a metal profile produce modifications in the interference pattern under coherent light illumination, and the speckle pattern changes in time.

Corrosion phenomena, then, can also be studied by the biospeckle technique. In this case, the corrosion deterioration of the sample is a long-time process. Experiments with copper, brass, and stainless steel, and copper with corrosion inhibitor plates with their surfaces polished and cleaned, was used as sample, and a speckle correlation technique was used to study the oxidation in real time.[14]

The oxidation process was reported to be evaluated using an experimental setup including a holographic register of the initial state,[14] showing that it is possible to evaluate the phenomenon on the surface of metal within 12 h. In that period, it was possible to distinguish, using decorrelation functions, the different types of behavior among four types of metals.

Corrosion grade was also monitored in iron samples by Begeman et al.[15] who developed a theoretical model, presented experimental results, and proposed the technique as a potential tool in nondestructive testing of metal corrosion. That work tried to cover a range of possibilities involving different surface roughness, besides the evaluation of the influence of reflectivity in the results. Corrosion was generated by exposition to an acid, and the correlation function was used for measurement in a backscattering geometry setup.

Another report evaluating the correlation of an electroerosion process in a metallic surface reinforced the ability of dynamic speckle to reproduce the micro-changes in the surfaceís profile, and was discussed in Chapter 3, followed by a theoretical model of the process.

9.5 SALT EFFLORESCENCE

Aging by salt crystallization is employed by scientists to investigate the decay in stone materials. However, stone damage can be reproduced very rapidly, and in some cases, only a few crystallization cycles are sufficient to fracture a stone specimen. The monitoring of salt efflorescence on stone surfaces, a problem suffered by some ancient monuments, has also been performed by following speckle changes with time.[16] A similar approach was also used to monitor water content in seeds.

P. Zanetta and M. Facchini[16] applied a local speckle correlation method to characterize the surface alteration mechanism in stones caused by salt efflorescence. They describe how local modifications of the surface microstructure can be adequately monitored by measuring the corresponding time evolution in the speckle image. The

correlation coefficient is proportional to the speed at which the water solution evapo-
rated and the salt efflorescence appeared at the surface.

Backscattering geometry is evidently required in this case as the objects are
opaque. A correlation measurement defined in Equation 9.2 was used to compare pairs
of images corresponding to different instances of the efflorescence phenomenon.

$$\rho = 1 - \frac{\left\langle (I_1 - I_2)^2 \right\rangle}{2m^2} \qquad (9.2)$$

The so-defined measurement of decorrelation ρ is expressed as the relation between
the squared difference between two images I_1 and I_2 and the mean value (m) of the
intensity. The angular brackets represent the averaging operation.

The slow response of the phenomenon requires hours to develop, and it is very
advantageous as it relaxes the time-resolution requirements. The results were depicted
as images showing the active areas where changes in the salt crystals modified the
correlation coefficient.

9.6 FOAMS STABILITY

Foams are widely present in our lives, with many characteristics based on their use or
their undesirable presence. Many times we can relate them with the pleasure linked
to a good chilled mug of beer, or a cup of hot espresso coffee accompanied by a
generous piece of cake with its nice covering of meringue foam. Foams are also present in shaving products, detergent solutions, fire extinguishers, and in many others
products. Therefore, the assessment of foams is an important topic in some related
industries, with the purpose of increasing stabilization or fast deterioration.

Foam is a coarse dispersion of a gas in a liquid. It is a colloidal in the sense that
the liquid films separating the gas cells are of colloidal thickness. Like emulsions,
foams are thermodynamically unstable, which means that foams will deteriorate
with time. The rate of deterioration of bubbles of foam depends of three physical
phenomena: drainage, coalescence, and disproportionation.

The dynamic speckle that occurs in the development of foams has a different
origin and features in comparison with the other studied examples. The activity here
is due to the change in size, rupture, and redistribution of bubbles.

One of the attributes that requires a special attention is its stability, which roughly
can be classified as unstable, with a transient form or as metastable, meaning a large
period of lifetime that can be from hours to days.

Among many possibilities to monitor the behavior of foam in time, the dynamic
speckle techniques provides an interesting alternative. For example, shaving foam
evolves slowly. So, it does not require a very high sampling rate and can be followed
by dynamic speckle techniques. One report found in the literature[17] provides an
evaluation of two different shaving foams under laser illumination using the inertia
moment values and adopting the backscattering approach to take the images.

Bubble explosion is a phenomenon that can also be observed using dynamic speckle
methods by the discontinuities in the time history of speckle pattern (THSP) image.
Figure 9.3 offers an image of the discontinuities that occurred in the foam of detergent.

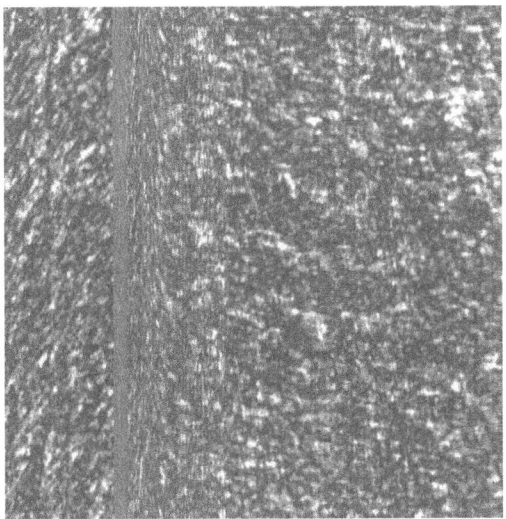

FIGURE 9.3 Time history of speckle pattern (THSP) image with some disrupting bubbles.

Experiments to evaluate the activity of foam as a function of time were successfully performed.[17,18] The experimental results show the time evolution of the samples, permit comparisons between different sample products, and show a potential applicability of the speckle technique to the assessment of this type of processes.

9.7 FURTHER APPLICATIONS

Some applications of dynamic speckle, such as to monitor the movement of particles[19] in a liquid, to measure the velocity of diffuse objects,[20] measure distances,[21] and identify inclusions in bulk, could also be included here to show the potential use of that technology. Cement hydration has also been reported as an interesting phenomenon that can be followed by using dynamic speckle techniques.[22]

9.8 CONCLUSIONS

A huge amount of nonbiological applications can be found to be possible using the dynamic speckle as a practical tool to help to solve metrological problems. Each demand requires a special design of illumination, and grabbing and assembling of the images, as well as a specific tool of analysis. The main advantage of the dynamic speckle as a powerful nondestructive technique that can be harnessed to applications requiring high sensitivity is its ability to acquire, without any contact, a high number of simultaneous data samples that are representative of the same physical phenomenon. Some drawbacks observed are the absence of a universal equipment and the restrictions in use related to vibration isolation and the requirement of darkness, as in its current state most of the equipment is restricted to metrological laboratories. Efforts to construct specific equipment and improve the capability to analyze the data with higher spatial and temporal resolution, as well as the characterization of the involved physical phenomena, should be pointed out as the main actions expected, besides the search for new applications.

REFERENCES

1. Arizaga, R., Trivi, M., and Rabal, H., Speckle time evolution characterization by the co-occurrence matrix analysis, *Opt. Laser Tech.*, 31(2), 163, 1999.
2. Passoni, I., Dai Pra, L. I., Rabal, H., Trivi, M., and Arizaga, R., Dynamic speckle processing using wavelets-based entropy, *Opt. Comm.*, 246, 219, 2005.
3. Fernández Limia, M. Mavilio Núñez, A., Rabal, H. J., and Trivi, M., Wavelet transform analysis of dynamic speckle patterns texture, *Appl. Opt.*, 41(32), 6745, 2002.
4. Passoni, L. I., Rabal, H. J., and Arizmendi, C. M., Characterizing dynamic speckle time series with the Hurst coefficient concept, *Fractals*, 12, 319, 2004.
5. Yamaguchi, I., Yokota, M., Ida, T., Sunaga, M., and Kobayashi, K., Monitoring of paint drying process by digital speckle correlation, *Opt. Rev.*, 14(6), 362, 2007.
6. Mavilio Nuñez, A., Fernandez Limia, M., Taño Lazo, M., Rabal, H. J., Arizaga, R., and Trivi, M., Characterization of dynamic speckles by the difference histogram method, *Opt. Eng.*, 46(5), 057005, 2007.
7. Federico, A. and Kaufmann, G. H., Evaluation of dynamic speckle activity using the empirical mode decomposition method, *Opt. Comm.*, 267, 287, 2006.
8. Federico, A. and Kaufmann, G. H., Multiscale analysis of the intensity fluctuation in a time series of dynamic speckle patterns, *Appl. Opt.*, 46(11), 1979, 2007.
9. Rabal, H.J., Cap, N., Trivi, M., Arizaga, R., Federico, A., Galizzi, G. E., and Kaufmann, G. H., Speckle activity images based on the spatial variance of the phase, *Appl. Opt.*, 45, 8733, 2006.
10. Braga, R. A., Silva, B. O., Rabelo, G., Costa, R. M., Enes, A. M., Cap, N., Rabal H. J., Arizaga, R., Trivi, M., and Horgan, G., Reliability of biospeckle image analysis, *Opt. Lasers Eng.*, 45, 390, 2007.
11. Amalvy, J. I.. Lasquibar, C. A., Arizaga, R., Rabal, H. J., and Trivi, M., Application of dynamic speckle interferometry to the drying of coatings, *Progr. Org. Coat.*, 42(1), 89, 2001.
12. Burnel, L., Brun, A., and Snabre, P., Microstructure movements study by dynamic speckle analysis, in *Speckle 06*, Slangen, P. and Cerrutti, C., Eds., *Proc. of SPIE*, 2006, 6341.
13. Puttappa, J., Joenathan, C., and Khorana, B., Speckle pattern analysis method and system, U.S. Patent 7.123.363.B2, 2006.
14. Muramatsu, M., Guedes, G. H., and Gaggioli, N. G., Speckle correlation used to study the oxidation process in real time, *Opt. Laser Tech.*, 26(3), 167, 1994.
15. Begemann, T.F., Gülker, G., Hinsch, K. D., and Wolff, K., Corrosion monitoring with speckle correlation, *Appl. Opt.*, 38(28), 5949, 1999.
16. Zanetta, P. and Facchini, M., Local correlation of laser speckle applied to the study of salt efflorescence on stone surfaces, *Opt. Comm.*, 104, 35, 1993.
17. Martínez, A., Ortiz, C., Arizaga, R., Rabal, H. J., and Trivi, M., Temporal evolution of speckle in foams, in *5th Iberoamerican Meeting on Optics and 8th Latin American Meeting on Optics, Lasers, and Their Applications*, Marcano, O. and Paz, J. L., Eds., *Proc. SPIE*, 2004, 1484.
18. Bandyopadhyay, R., Gittings, A. S., Suh, S. S., Dixon P. K., and Durian, D. J., Speckle visibility spectroscopy: a tool to study time varying dynamics, *Rev. Scientific Instruments*, 76, 093110, 2005.

19. Mizukami, A. and Muramatsu, M., Correlação de speckle dinâmico produzido por espalhamento de luz por partículas num meio líquido, in *XVIII Encontro Nacional de Física da Matéria Condensada*, Caxambu, 1997.
20. Asakura, T. and Takai, N., Dynamic laser speckles and their applications to velocity measurement of the diffuse object, *Appl. Phys.*, 25, 179, 1981.
21. Semenov, D. V., Nippolainen, E., and Kamshilin, A. A., Accuracy and resolution of a dynamic-speckle profilometer, *Appl. Opt.*, 45, 411, 2006.
22. Gorsky, M.P, Maksimyak, A. P., and Maksimyak, P. P., Study of speckle-field dynamics scattered by surface of concrete during congelation, in *Advanced Topics in Optoelectronics, Microelectronics, and Nanoelectronics III,* Lancu, O. et al., Eds., *Proc. SPIE*, 2007, 66350E.

10 Didactic Dynamic Speckle Software

Marlon Marcon and Roberto A. Braga Jr.

CONTENTS

10.1 INTRODUCTION

The challenge involved in starting to use speckle images can be so much of an obstacle that newcomers in the area could be discouraged by the amount of time required. In order to induce a soft start or to create a net of programs that can be constructed collectively, we decided to gather together in this book some macros, or collection of instructions, to be used in a free platform, in this case the ImageJ.

The macros constructed were Brier's Contrast, Generalized Differences (GD), Fujii Method, and Inertia Moment (IM). They are thoroughly and openly presented so that those who want to create their own set of programs, and to start getting some results, may do so, and also so that any possible improvements in the programs can be shared and enjoyed by all.

The chapter begins by helping to generate a collection of speckle images using extremely basic equipment to illuminate and capture data, and further to create stacks with the images and use them.

10.2 HOW TO CREATE YOUR OWN DYNAMIC
SPECKLE COLLECTION OF IMAGES

An optical laboratory with an optical table supported by a noise vibration suppression system, laser, CCD, microcomputer, and many optical devices such as lenses, post-holders, mirrors, and so on, gives you the best installation for grabbing your images. However, if this is not yet possible, you can use a diode laser found in the pens known as laser pointers. You should be aware of the low quality of that beam, but it is enough for a beginning with dynamic speckle.

One suggestion for maximizing the use of your pen is to power it with a power source commonly found in many electronic devices instead of with batteries. You should be careful to choose one with the correct voltage level, as you will need to have the laser source working for a long time without consuming all the batteries you have.

You will also need a beam splitter, a lens, depending on the area of illumination you want, and finally a digital camera. The design of an alternative setup can be seen in Figure 10.1, where the table and the arm can be constructed by yourself or adapted from applications such as a photography set.

Using the camera, you can grab a film or a set of frames. It is very important to know that any noise related to vibration should not be introduced in the system during the process of image collection. In this case, a camera with a remote trigger will help your work, and eliminate the possibility of introducing vibrations in the setup during each touch in the camera.

You can choose many samples to start your experience, anything from wet seeds all the way through to the classic example of a coin just painted. The images of the samples could be focused or not by the camera, which will produce different results expressing the same phenomenon.

In using of the camera you will be able to change the speed that, in some cases, will produce clear differences, such as in the Brier's Contrast results.

The user should also consider the influence of other light sources during the illumination process, as the laser pointer emits some very low light. When the application requires the absence of any light source, total darkness only is sufficient.

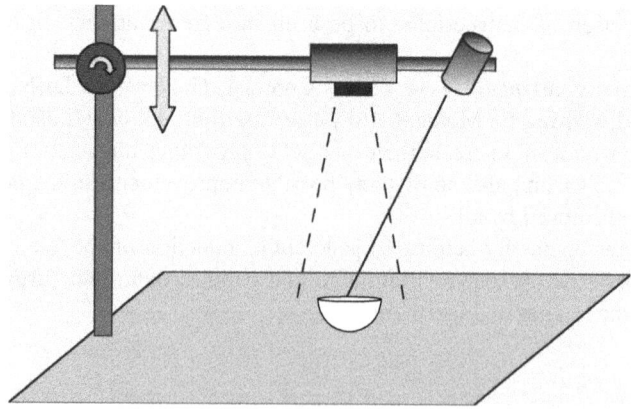

FIGURE 10.1 Basic design of illumination and image assembling.

FIGURE 10.2 The main ImageJ screen captured by Ubuntu© version 7.10 Linux system.

10.3 IMAGEJ: HOW AND WHY

ImageJ is a public domain image processing program written in Java. It means that it runs on any computer which has Java Runtime Environment installed (e.g., Linux, Windows, MacOSX, Solaris, even PDAs).[1] There are downloadable distributions available as well as an online applet: http://rsb.info.nih.gov/ij/index.html. Figure 10.2 shows the ImageJ's main screen.

ImageJ can display, edit, analyze, process, save, and print 8-bit, 16-bit, and 32-bit images. It can read many image formats including TIFF, GIF, JPEG, BMP, DICOM, FITS, and "raw," and it supports "stacks," a series of similar 2-D images that can be manipulated together and share a single window.[2-3]

The functions provided by ImageJ's built-in commands can be extended by user-written code in the form of macros and plug-ins. Macros are an easy way to automate a collection of commands assembled in a text file. The macro code can be created and modified in the built-in editor as well as in any text editor. The ImageJ macro language has a set of built-in functions and control structures which can be used to call other built-in commands and macros.[4] Otherwise plug-ins are much more powerful than macros; they are implemented as Java classes, which means that you can use all features of the Java language,[2] though it is not so trivial as the friendly use of macros to build the programs. More detailed references about the ImageJ can be found in some books,[5,6] in online references[2,7] and at the ImageJ Information and Documentation Portal: http://imagejdocu.tudor.lu/.

The use of a high-level platform makes the startup easier, and the facility presented by a free software helps the use by a greater number of students, and beginning and senior researchers, who can certainly introduce improvements and share their experiences.

10.4 HOW TO CREATE A STACK OF IMAGES

A stack of biospeckle images can be created easily once you have a collection of frames of speckles. There are many different ways to create it in the ImageJ platform, depending on the data type you have. If you have a film, you can convert it to a stack using the menu "File/Import/Video" with the necessary plug-ins indicated by the program. Another way is using an animated GIF, which is split into frames and converted to a stack by the menu File/Import/Animated GIF . . . . Figure 10.3 shows how to access to the Import menu.

If you have a collection of images in the same folder, you can use the menu File/Import/Image Sequence and choose the first image of your sequence; the ImageJ will open them,

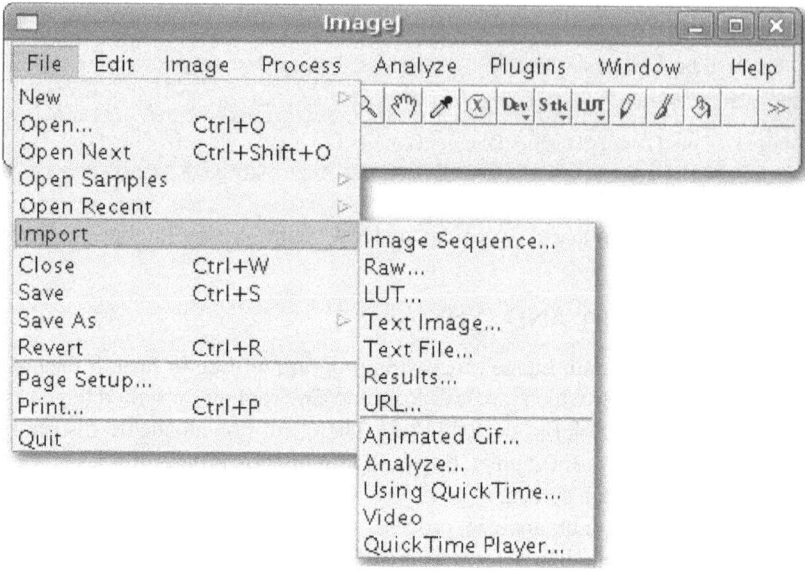

FIGURE 10.3 ImageJ's Import menu and its options.

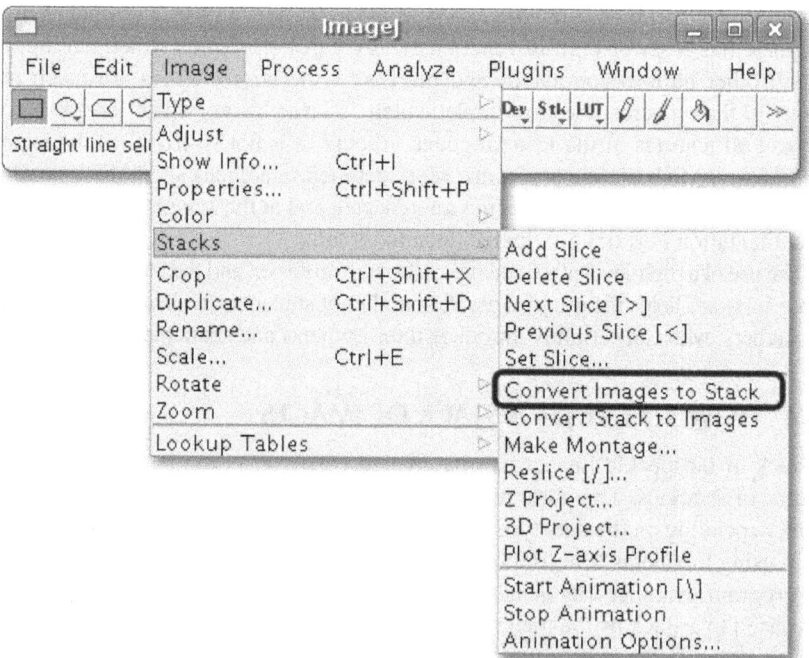

FIGURE 10.4 ImageJ menu to convert images to a stack.

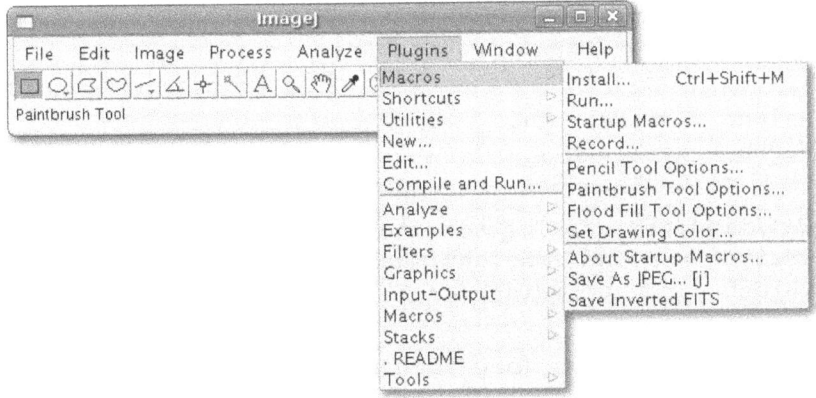

FIGURE 10.5 Plugins' menu in ImageJ.

sorted by file name (see Figure 10.3). Another way is to open all the images and select the menu Image/Stacks/Convert Images to Stack, as can be seen in Figure 10.4.

10.5 HOW TO EDIT AND INSTALL A MACRO INSIDE IMAGEJ

The analysis of your images can be started now. In the next sections you will be presented instructions of ImageJ linked together in order to create macros to implement some biospeckle analysis. To copy the macros proposed in the next sections, you can use any text edit program, or the edit environment inside ImageJ, using the option Plugins/Edit or Plugins/New. You can save your macro to any place on your computer, and when you run the macro it will automatically ask you where it is. Figure 10.5 shows where you will find these operations in ImageJ.

10.6 BRIERS' CONTRAST MACRO (LASCA)

Briers' Contrast macro allows you to evaluate the contrast of any image; therefore, you have a tool to enhance the edges of an object. The concept in a speckle image is to screen areas distinguishing the contrast of the image. You will need just one image to run the macro and get a new image with Briers' Contrast. The speed of the process, the clear difference between activity in close areas, and the exposure time will be determining factors in your results.

For example, if you have a high activity phenomenon, and if you grab it at low speed, you will certainly lose information.

The macro is complete in the following lines, and after editing and saving it you can run it, but remember that you must open the image before running the macro. It is important to remember that the program will ask you the size of the mask, which means that the Briers' Contrast will be implemented evaluating a window of pixels. This also means that the size of the generated image will be smaller losing spatial resolution. The window size default value is 7. Each image format supported by ImageJ, and each size will be accepted in this macro, and in addition, any image, speckle pattern or not, can be used.

```
//-----------------------------------------------------------
//Macro ImageJ-Briers' contrast
//-----------------------------------------------------------
imHeight = getHeight();
imWidth = getWidth();
mask = getNumber("Set the size of the mask...", 7);

//Defining the size of new image
newHeight = round(imHeight/mask);
newWidth = round(imWidth/mask);

//New image array
newResult = newArray(newWidth*newHeight);

//The Method
for(l = 0; l < newWidth; l++){
    for(k=0; k < newHeight; k++){
        aux = newArray(mask*mask);
        for(i=0; i < mask; i++){
            for(j = 0; j < mask; j++){
                aux[j*mask + i] = getPixel((l*mask + i),
                  (k*mask + j));
            }
        }
        newResult[l*newHeight + k] = Briers(aux);
    }
}
//Showing result image
newImage("Briers...", "8-bit Color White", newWidth ,
newHeight, 1);

for(i = 0; i < newWidth; i++){
    for(j = 0; j < newHeight; j++){
        putPixel(i, j, newResult[i*newHeight + j]);
    }
}
//Function to determine the Brier's Contrast of an array,
//coef = 1 - (std/mean)
function Briers(arr){
sum = 0;
sumSqr = 0;
    for(i = 0; i < arr.length; i++){
        sum= sum + arr[i];
        sumSqr = sumSqr + (arr[i]*arr[i]);
    }
    mean = sum/arr.length;                          //Mean
    variance = (sumSqr/arr.length)-(mean*mean); //Variance
```

```
    std = sqrt(variance);              //Standard deviation
    return (round((1 - (std/(mean + 1)))*255));
}
// End of Briers' Contrast macro
```

10.7 GENERALIZED DIFFERENCES AND FUJII METHOD

The GD and Fujii methods are also presented here in accordance with the algorithms discussed in Chapter 5. In GD algorithm, a window is shifted throughout the stack of frames and produces a subtraction of only a range of five frames (default value, defined in GD Macro) in a sequence followed by summation of the absolute value of the differences.

We made two different approaches to develop GD algorithm: in the first, a window with a set of frames is the range where the images are subtracted and the results summed; the window after the comparisons is shifted while it fits in stack (stkSize–window). As can be seen, it will be easy to change the size of the window at will, by changing the number of subtractions.

In the second, each image was compared (subtracted) with the images that follow until the end of the stack.

Remember that you have, in both cases of Fujii and GD, to open a stack of images before starting the application, as mentioned in Section 10.4.

10.7.1 GD Method Macro

```
//-----------------------------------------------------------
//Macro ImageJ—Generalized Difference Method (GD)
//-----------------------------------------------------------

//Size of the stack
stkSize = nSlices;

imHeight = getHeight();
imWidth = getWidth();

//size of the window
Window = getNumber("Set the size of the window...", 5);

//Array of the Image result
arr = newArray(imHeight*imWidth);

//GD Method routine 1st approach
for(i = 1; i <= stkSize - window; i++){
    setSlice(i);
    pix1 = newArray(imHeight*imWidth);
    for(y = 0; y < imHeight; y++ ){
        for(x = 0; x < imWidth; x++){
            pix1[y*imWidth + x] = getPixel(x,y);
        }
    }
```

```
    for(j = 1; j <= window; j++) {
        setSlice(i+j);
        for(y = 0; y < imHeight; y++ ) {
            for(x = 0; x < imWidth; x++) {
                pixel1 = pix1[y*imWidth + x];
                pixel2 = getPixel(x,y);
                arr[y*imWidth + x] = arr[y*imWidth
                    + x] + abs(pixel1 - pixel2);
            }
        }
    }
}
// End of routine 1st approach
min = 256;
max = -1;
//Discovering the maximum and minimum value to normalize
the result
for(i = 0; i < imWidth*imHeight; i++) {
    min = minOf(min,arr[i]);
    max = maxOf(max,arr[i]);
}
//Plotting the image
newImage("GD...", "8-bit White", imWidth , imHeight, 1);
for (y = 0; y < imHeight; y++) {
    for (x = 0; x < imWidth; x++) {
        putPixel(x,y,round((arr[y*imWidth+x]/
            (max- min))*255));
    }
}
// End of GD macro
```

To execute the second approach, you may change the code of GD method routine by the code that follows:

```
//GD Method routine 2nd approach
for(i = 1; i <= stkSize; i++) {
    setSlice(i);
    pix1 = newArray(imHeight*imWidth);
    for(y = 0; y < imHeight; y++ ) {
        for(x = 0; x < imWidth; x++) {
            pix1[y*imWidth + x] = getPixel(x,y);
        }
    }
    for(j = 1; (j <= stkSize) && ((i + j) <= stkSize);
        j++) {
        setSlice(i+j);
```

```
      for(y = 0; y < imHeight; y++ ){
          for(x = 0; x < imWidth; x++){
                pixel1 = pix1[y*imWidth + x];
                pixel2 = getPixel(x,y);
                arr[y*imWidth + x] = arr[y*imWidth + x] +
                  abs(pixel1 - pixel2);
                }
          }
      }
}
// End of routine 2nd approach
```

10.7.2 Fujii Method Macro

Fujii method is implemented in the macro that follows as presented in Chapter 5.

```
//--------------------------------------------------------
//Macro ImageJ—Fujii Method
//--------------------------------------------------------
imHeight = getHeight();
imWidth = getWidth();

//Array of the Image result
arr = newArray(imHeight*imWidth);

//Number of images of ImageJ's stack
numImages = nSlices;

//Method
for(i = 1; i < numImages; i++){
      setSlice(i);
      pix1 = newArray(imHeight*imWidth);
      cont = 0;
      for(y = 0; y < imHeight; y++ ){
              for(x = 0; x < imWidth; x++){
                  pix1[y*imWidth + x] = getPixel(x,y);
      }
    }
    setSlice(i+1);
    for(y = 0; y < imHeight; y++ ){
          for(x = 0; x < imWidth; x++){
                pixel1 = pix1[y*imWidth + x];
                pixel2 = getPixel(x,y);
                arr[y*imWidth + x] = arr[y*imWidth + x] +
                abs((pixel1 - pixel2)/(pixel1 + pixel2));
                }
          }
      }
}
```

```
//Definition of the maximum and minimum values (for data
normalization)

min = 256;
max = -1;
for(i = 0; i < imWidth*imHeight; i++){
        min = minOf(min,arr[i]);
        max = maxOf(max,arr[i]);
}

//Showing the image
newImage("Fujii...", "8-bit White", imWidth ,
imHeight, 1);
cont = 0;
for (y = 0; y < imHeight; y++){
        for (x = 0; x < imWidth; x++){
                putPixel(x,y,round((arr[y*imWidth+x]/
                (max-min+1))*255));
        }
}
// End of Fujii macro
```

10.8 INERTIA MOMENT

The IM macro will offer you the possibility to create and to plot a THSP of a stack, to create and to plot the Co-occurrence Matrix, and finally to calculate the numerical value of IM.

You should take into account that the absolute value of the IM alone does not have an intrinsic meaning otherwise, it is useful to provide a relative activity measure to be compared with other IM values in a set of related experiments involving the same conditions.

```
//-----------------------------------------------------
//Macro ImageJ—Inertia Moment
//-----------------------------------------------------

//Size of the stack
stkSize = nSlices;

imHeight = getHeight();
imWidth = getWidth();

//-----------------------------------------------------
//THSP Construction
//-----------------------------------------------------

stsHeight=512;
if(imWidth < 512){
```

```
      stsHeight = imWidth;
}
//Array of the Image result
arr = newArray(stsHeight*stkSize);

//Mounting the STS
min = round((imWidth - stsHeight)/2);
max = min + stsHeight;

for(i = 0; i < stkSize; i++){
      setSlice(i+1);
      for(x = min; x < max; x++){
             arr[(x-min)*stkSize + i] = getPixel(x,round
                (imHeight/2));
      }
}

//Plotting the image
newImage("STS...", "8-bit White", stkSize , stsHeight, 1);
cont = 0;
for (i = 0; i < stkSize; i++){
       for (x = 0; x < stsHeight; x++){
              putPixel(i,x,round(arr[x*stkSize+i]));
       }
}

//-----------------------------------------------------
//Occurrence Matrix Construction
//-----------------------------------------------------

moc = newArray(256*256);
for (i = 1; i < stkSize; i++){
       for (x = 0; x < stsHeight; x++){
              moc[arr[x*stkSize+i-1]*256 + arr
                [x*stkSize+i]] = moc[arr[x*stkSize+i-1]
                *256 + arr[x*stkSize+i]]+1;
       }
}
sumArray = newArray(256);
for (i = 0; i < 256; i++){
       for (x = 0; x < 256; x++){
              sumArray[x] = sumArray[x] + moc[x*256+i];
       }
}
```

```
//Inertia Moment's Matrix (Co-occurrence matrix)
for (i = 0; i < 256; i++){
        for (x = 0; x < 256; x++){
                moc[x*256+i] = moc[x*256+i]/
                (sumArray[x]+1);
        }
}
//Plotting the image
newImage("MOC...", "8-bit White", 256 , 256, 1);

for (y = 0; y < 256; y++){
        for (x = 0; x < 256; x++){
                putPixel(x,y,round(moc[y*256+x]*256));
        }
}

//-------------------------------------------------------
//Inertia Moment Calculation
//-------------------------------------------------------

IM = 0;
for(y = 0; y < 256; y++){
        for(x = 0; x < 256; x++){
                IM = IM + moc[x*256 + y]*pow((x-y),2);
        }
}

//IM value
print("Inertia Moment = " + (IM/256));

// End of Inertia Moment macro
```

10.9 CONCLUSIONS

We hope the macros presented in this chapter will help you start dealing with dynamic speckle phenomena, and to develop your own works. The macros cover four approaches to deal with image analysis and to get useful values related to activity from speckle patterns. The open source platform and the short macros were our strategies to share with you our basic tools and to give you experience with this interesting phenomenon.

REFERENCES

1. Rasban, W., ImageJ [Online], Available: http://rsb.info.nih.gov/ij/.
2. Bailer, W., Writing ImageJ PlugIns: A Tutorial, [Online] available: http://www.imagingbook.com/fileadmin/goodies/ijtutorial/tutorial171.pdf.
3. Hessman, F. V. and Modrow, E., An Introduction to Astronomical Image Processing with ImageJ [Online], available: http://www.uni-sw.gwdg.de/~hessman/ImageJ/Book/.

4. McEvoy, F. J., An application of image processing techniques in computed tomography image analysis, *Vet. Radiol. Ultras.*, 48(6), 528, 2007.
5. Burger, W. and Burge, M. J., *Digital Image Processing with Java and ImageJ*, Springer, New York, 2008, p. 566.
6. Berry, E., *A Practical Approach to Medical Image Processing*, Taylor & Francis, Boca Raton, FL, 2007, p. 304.
7. Collins, T., Online Manual for the MBF-ImageJ Collection [Online], available: http://www.macbiophotonics.ca/imagej/.

11 Perspectives

Hector Jorge Rabal and
Roberto A. Braga Jr.

CONTENTS

11.1 INTRODUCTION

In this book we intended to organize the main theoretical concepts of dynamic speckle, the techniques adopted to work with the phenomenon, and various other applications. It should be clear that in such a technological area it is impossible to cover the subject completely. Nevertheless, while attempting to review as much as possible, it was observed that the application areas demand a deep comprehension of the phenomena involving the interaction of the laser and the matter under illumination, in some cases calling for special and specific accounts involving a multidisciplinary evaluation.

The aim of this book, therefore, is not to cover the subject completely, but to inform people about the technology and its challenges. Upcoming research will continue to help science to understand better the complex interaction between coherent light and matter, biological and otherwise. In this way, many efforts to develop reliable technologies will be made and certainly will add to the development of equipment and systems in many areas requiring a nondestructive technique.

We pointed out in Chapter 10 some of our dreams, hopes, and intentions to continue development in this area. Some comments are necessarily speculative with different degrees of plausibility, and try to pass throughout the book discussing some interesting approaches.

11.2 KNOWLEDGE OF INNER PHENOMENA

As in Chapter 2, on the origins and characteristics of dynamic speckle, something that will hopefully happen is an increase in the knowledge of the inner dynamics of the phenomena that give rise to the dynamic behavior. As inner dynamics of the samples are better known, a better insight can be obtained in controlled experiments and simulations to assess how these dynamics show up in the speckle evolution. The possibilities of the use of speckle in other tests would benefit from this knowledge.

Some other phenomena will probably be included as causes that have not been explored in depth. Mitochondria[1] are known to exhibit rotatory power, an optical effect that rotates the polarization plane, which in this case depends on the action potential and should manifest as an activity, albeit weak.

When an object moves in the path of an illuminating beam and projects a shadow on a static diffuser, it covers and uncovers different regions of the diffuser. The contributions of scattered waves in the newly uncovered regions and the lack of contributions from the hidden ones should manifest as a dynamic speckle, as can be easily verified and simulated in Figure 11.1.

As in Chapter 3, theoretical and numerical models will most probably evolve to include more and more realistic situations. Recent use of the copula will hopefully evolve to include simulation of activity images and to simulate situations that emulate experimentally obtained power spectra.

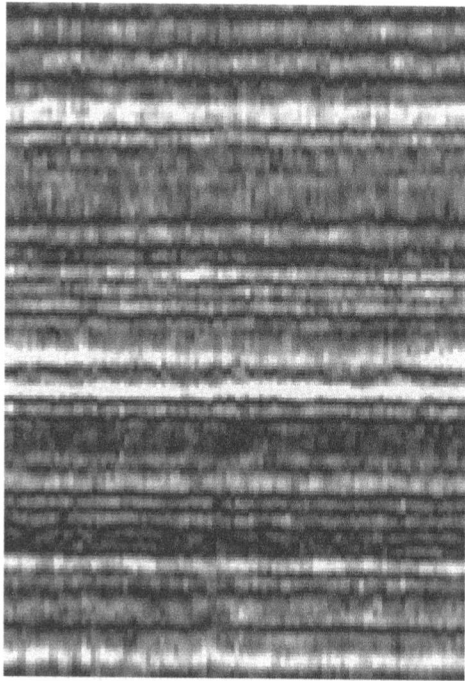

FIGURE 11.1 Simulated THSP (Time History Speckle Pattern) where the origin of the activity is in the moving shadows of an object in front of a diffuser.

The tools described in Chapter 4 will probably be increased and hopefully a better insight in the interpretation of the results will arise from the multidisciplinary interaction with different experts. In particular, efforts to distinguish the contributions of the different causes, even in a first coarse way, are improvements that would be welcome.

The discrimination of the contributions to the activity of two different components, such as semen and its diluents, is another problem that remains to be satisfactorily solved.

11.3 METHODOLOGIES

Chapters 4, 5, 6, and 7, for example, covered many methodologies to reproduce the light-matter interactions and to get useful information from them. On the other hand, the proposition of new methods of analysis will certainly help in the increase of reliable and specific information. New statistical approaches and image analysis are welcome, and will be based on the development of the equipment currently used to deal with signal and data, such as computers, frame-grabbers, and cameras.

In addition to software and procedures, it is necessary to unfold the possibility to achieve the results using new designs of equipment as well as using new equipment.

11.4 EQUIPMENT

Fast and low-cost dedicated devices, such as optical sensors with high spatial and temporal resolution will be continuously demanded and expected. Cheap detectors, such as those employed in an optical mouse detectors and diode lasers could permit the implementation of instruments dedicated to some specific operations accessible for massive production and use. These are interesting devices that could be fast adopted or adapted. The usage of optical mouses connected to a notebook doing many sorts of evaluations should not be regarded as a dream; instead, it should be seriously considered. Some measurements using very few detectors have shown to give reliable results.[2]

In the same way, the development of fast and more sensitive optical sensors will help the usage in areas were the activity monitored is faster than the accessible one with current devices that filter some information.

Computers are one of the most basic equipment, though responsible for creating conditions to many developments in this area. Certainly, they will still be the difference in new works, by helping the fast acquirement of signal, analyzing the data in a quasi-real time, or being the environment to new virtual systems to simulate or evaluate the data in an expertise way.

In addition, some of the applications presented in Chapter 8 demand the construction of dedicated systems or equipment to conduct specific analysis, such as in semen evaluation. Nowadays, it is possible to find at least two commercial equipment to perfusion analysis in human beings and paint drying evaluation, showing that it is already reliable to cross the bridge between science and technology in this area.

Activity images will continue to improve as a consequence of increase in the performance of the cameras used to register the images. Better intensity quantization, spatial and temporal resolution along with corresponding higher sensitivities, will help to improve quality and speed of acquisition of these images permitting access to other phenomena. The use of image intensifiers for these types of phenomena seems to be practically unexplored.

There are still features in the activity images of unknown origin that appear in the experimental results when some techniques are used, but not with some others.

The applications in the agriculture and veterinary fields are expected to increase as the involved phenomena fit into the sampling frequency window of CCD cameras for one side and constitute very interesting economic possibilities.

11.5   MODELS AND EXPERT PROGRAMS

Chapter 3 presented some models to simulate the dynamic speckle behavior; nevertheless, we should recognize the necessity of continuous development of theoretical and numerical models increasing the complexity of the simulation of the interaction of light-matter. This would help to a better understanding of the phenomena, especially by decoding all the influences that are present in the formation of speckle interference patterns.

Expert programs will certainly help during the data analysis, by producing the automated selectivity of the many features that could be part of a complex result. Chapter 7 discussed fuzzy algorithms unfolding the use of other algorithms such as neural networks.

Chapter 5 presented an approach, receiving operator characteristic (ROC), to deal with dichotomy results also unfolding the area to other optimization methodologies.

11.6   VISUALIZATION INSIDE THE TISSUES

The investigations of the phenomena inside tissues are presented in Chapter 8 as a current challenge, with some good results. Nevertheless, a huge effort should be made in this direction, since much useful information is probably being lost with the current approaches, and methodologies to separate the areas or layers of interest with more accuracy are still required. A discussion placed in Chapter 8 was about the difficulties and facilities to obtain information from the interior of tissues. One case that represents the facility is the flux identification (perfusion). On the other hand, the identification of areas not related to any flux of materials is still a challenge, for instance, the ability to distinguish parasites in seeds allowing the classification and selection of them in an automated way.

The use of the frequency domain to analyze the differences is presented as a potential approach. Some other suggestions[3,4] showed the conditions to obtain information from layers with activity inside a nonactive material. However, could the opposite be possible? One of the works presented a first step to solve this challenge,[5] because it was possible to identify high activity in the surface of a damaged apple. The bruising process is not due to a flux of materials; so, it can be observed that the same tissue can be identified in high and low activity. The challenge is now to achieve a low activity area inside the same tissue with high activity, and as deep as possible.

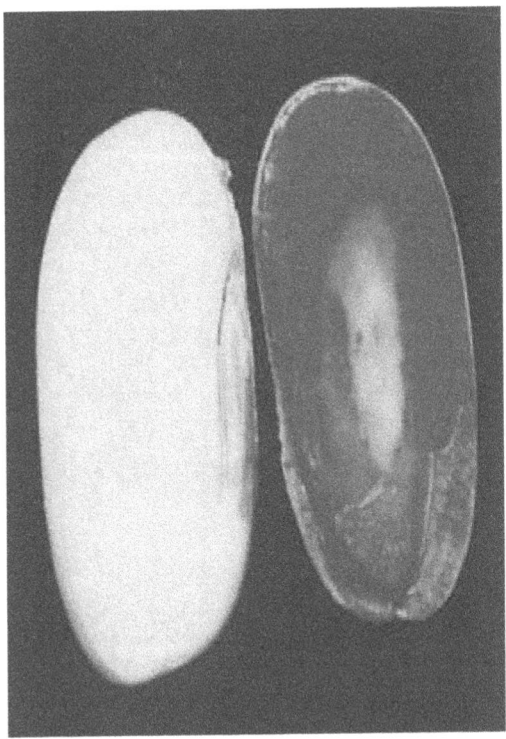

FIGURE 11.2 Internal portions of bean seed's cotyledon, one without any product and the other with tetrazolium salt identifying a symmetric damage in white.

Figure 11.2 shows the challenge to obtain information from damage inside a seed with high activity. A seed with symmetrical damage was analyzed, and one cotyledon received a salt to color the alive and dead areas. The results using Generalized Differences (GD) and Fujii methods are shown in Figure 11.3 and it is possible to see a trace of the damaged tissue, but still without clear definition.

The identification of damaged areas in seeds is a very important issue and new works intending to obtain better results should be done. For this purpose, the frequency approach seems to be an interesting way.

In this case, as the areas can be identified, expert systems will be necessary to classify the relevance of the damage with respect to location and size.

11.7 POLARIZATION

The polarization changes in interaction with active tissue. Collagen fibers, muscle fibers, some cells in the retina, and many others show birefringence. The information provided by the polarization state changes will probably be combined with the one obtained with the dynamic speckle in a more complex and interesting phenomenon. Laser light scattered from biological tissues produces randomly polarized speckle patterns[6] and the two orthogonally polarized components are incoherently added with

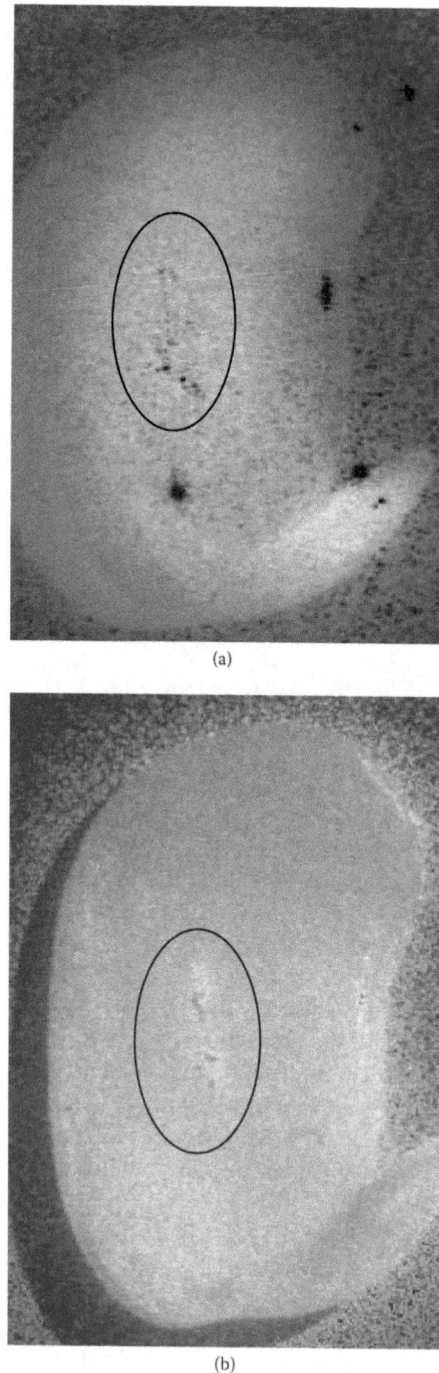

(a)

(b)

FIGURE 11.3 Images of (a) Fujii and (b) Generalized Differences results with the damage assigned.

each other in ordinary quadratic detection. Single speckle patterns arising from biological tissues that were randomly polarized can be considered as the incoherent combination of two independent speckle patterns and the resulting contrast is diminished. Speckle patterns originating from biological tissues exhibit fast decorrelation and the polarization state is randomized. Meanwhile, the instantaneous polarization in speckle phenomenon can be expected to occur, which would be dynamic if the sample also is. That is, it can be expected that in each point of the observed pattern many wavelets would be superposed with instantaneously different polarization states and phases. The phenomenon would not be observable without careful processing.

There exist, although quite impractical in their current state, register media with polarization memory, such as those involving Weigert effect,[7] which have been used to process static speckle patterns and holograms,[8] showing that the storage of such patterns is in principle feasible.

In order to clarify the idea with a very coarse idealized example, it can be pointed out that if the image of a sample exhibits a region where polarization remains unchanged in time (although occasional speckle is boiling there), and there is another region where polarization is "boiling," a register with polarization memory could be used to construct polarization activity images by using, for example, adequate adaptations of LASCA or other blur-based algorithms.

The use of polarization measures along with speckle information, such as polarized speckle size[9] and the distribution of polarization singularities,[10] have been recently proposed for medical applications, but dynamic properties have not received much attention.

11.8 RADIATION IN OTHER FREQUENCY RANGES

There are regions in the electromagnetic spectrum where emitters and detectors already exist but are very expensive, such as the terahertz region[11,12] corresponding to the submillimeter wavelength range between 1 mm (high-frequency edge of the microwave band) and 100 μm (long-wavelength edge of far-infrared light), which permits access to phenomena occurring at different depths, provided that the material has no high water content.

Ultrasound waves are known to show speckle phenomena; in addition, medical sonography images show the speckle phenomenon, and in some cases this speckle is dynamic. Directional ultrasound emitters (UASERs)[13] with certain analogies to optical lasers are also under development, and their use can be expected to be adopted in both medical and industrial applications.

The "many to one" ambiguity of the measures of speckle activity could be alleviated by using several wavelengths in the illumination. The combined use of more than one wavelength has not been fully explored yet, and additional information is needed.

11.9 HETERODYNE DETECTION

The recently reported use of heterodyne detection and holographic reconstruction is a new and fresh approach that has been providing very interesting results and will probably be another productive line of research.

11.10 STANDARDIZATION

The necessity of creating a pattern or collection of protocols to conduct analyses should be one of our efforts. An example could be finding a way to present the activity by frequency or by an adimensional number.

Polarization may or may not become the standard in achieving images. Establishing a protocol is relevant to the future of this field and should be defined in further discussions.

11.11 NEW APPLICATIONS

It seems clear that dynamic speckle applications, only touched in a light way here, look endless at this point. Many areas need or could make good use of such optical metrology based on a nondestructive procedure under contactless and nonperturbative actions in the samples, in some cases presented as a robust and automated tool. Mainly in biological areas, the dynamic speckle presents itself as a powerful tool to identify areas using image techniques, or to quantify activity levels using numerical methods.

An interesting challenge is the study of foam during its extinguishing process, that happens for example, in espresso beverage. The intense changes of the refractive index produced by the evaporation undermine the information from the foam that only survives in a hot beverage. Figure 11.4 shows this challenge. Notice that the foam in espresso plays a special role in the taste of the beverage; its stability is desirable and is not affected by the addition of sugar grains or drops as sweeteners.

In addition, optical singularities,[14,15] consisting of points where intensity is zero and consequently the phase is not defined, are present in speckle patterns. These points can be identified and located with extremely high precision[16,17] in static and quasi-static situations. Nanometer precision has already been achieved. The use

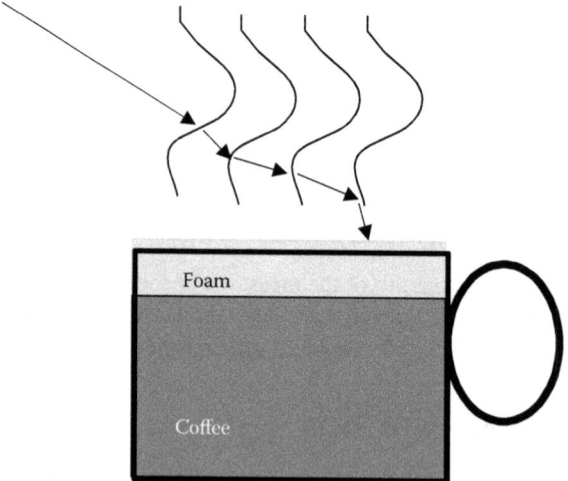

FIGURE 11.4 Schematic challenge of coffee foam measure with dynamic speckle.

of this vortex in dynamic speckle has practically not been developed and unfolds interesting possibilities, particularly in the measurement of very slow phenomena.

As the experimental setup for the simplest applications is not very expensive, the involved phenomena present a very interesting perspective for educational purposes, as experiments can be performed by students with high quality results, and they can be allowed to test their new ideas.

REFERENCES

1. Tychinsky, V., Kretushev, A., and Vishenskaja, T., Mitochondria optical parameters are dependent on their energy state: A new electrooptical effect, *Eur. Biophys. J.*, 33, 700, 2004.
2. Braga, R. A. et al., Reliability of biospeckle image analysis, *Opt. Lasers Eng.*, 45, 390, 2007.
3. Gonik, M. M., Mishin, A. B., and Zimnyakov, D. A., Visualization of blood microcirculation parameters in human tissues by time-integrated dynamic speckles analysis, *Ann. NY Acad. Sci.*, 972, 325, 2002.
4. Nothdurft, R. and Yao, G., Subsurface imaging obscured subsurface inhomogeneity using laser speckle, *Opt. Expr.*, 13(25), 10034, 2005.
5. Pajuelo, M. et al., Bio-speckle assessment of bruising in fruits, *Opt. Lasers Eng.*, 40, 13, 2003.
6. Cheng, H. et al., Efficient characterization of regional mesenteric blood flow by use of laser speckle imaging, *Appl. Opt.*, 42(28), 57, 2003.
7. Weigert, F., Uber Einen Neuen Effect der Strahling in Lichtempfindlichen Schichten, *Verh. Dtsch. Phys. Ges.*, 21, 479, 1919.
8. Rabal, H., Interference fringes produced by anisotropic speckle patterns, *Opt. Commun.*, 58(2), 69, 1986.
9. Piederriere, Y. et al., Speckle and polarization for biomedical applications, *Proc. SPIE*, 6341, 634106, 2006.
10. Angelsky, O.V. et al., Polarization singularities of the object field of skin surface, *J. Phys. D: Appl. Phys.*, 39, 3547, 2006.
11. Arnone, D. D. et al., Application of terahertz (THz) technology to medical imaging, in *Proc. SPIE Terahertz Spectroscopy Applications II*, International Society for Optical Engineering: Bellingham, 1999, p. 209.
12. Sherwin, M., Terahertz power, *NATURE*, 420, 2002, http://www.nature.com/nature 131.
13. Weaver, R. L. and Lobkis, O. I., Ultrasonics without a source: Thermal fluctuation correlations at MHz Frequencies, *Phys. Rev. Lett.*, 87, 134301, 2001.
14. Nye, J. and Berry, M., Dislocations in wave trains, *Proc. R. Soc. Lond. A.*, 336, 165, 1974.
15. Berry, M. V. and Dennis, M. R., Phase singularities in isotropic random waves, *Proc. R. Soc. Lond. A*, 456, 2059, 2000.
16. Wang, W. et al., Optical vortex metrology based on the core structures of phase singularities in Laguerre–Gauss transform of a speckle pattern, *Opt. Expr.*, 14(22),10195, 2006.
17. Wang, W. et al., Optical vortex metrology for nanometric speckle displacement measurement, *Opt. Expr.*, 14(1), 127, 2006.

Index

Milton Keynes UK
Ingram Content Group UK Ltd.
UKHW040445071024
449327UK00020B/1001